ADVANCES IN BIOCHEMICAL ENGINEERING

Volume 15

Managing Editor: A. Fiechter

With 39 Figures

Springer-Verlag Berlin Heidelberg GmbH 1980

ISBN 978-3-662-15408-3 ISBN 978-3-540-38513-4 (eBook)
DOI 10.1007/978-3-540-38513-4

Originally published by Springer-Verlag Berlin Heidelberg New York in 1980
Softcover reprint of the hardcover 1st edition 1980
Library of Congress Catalog Card Number 72-152360

2152/3140-543210

Managing Editor

Editorial Board

Professor Dr. A. Fiechter
Eidgenössische Technische Hochschule
Hönggerberg
CH-8093 Zürich

Contents

The Loop Reactor for Cultivating Yeast on n-Paraffin Substrate

Reinhold Seipenbusch
VEBA-Chemie AG, Werksgruppe Herne
D-4690 Herne 2, West Germany

Heinz Blenke
Institut für Chemische Verfahrenstechnik
Universität Stuttgart, D-7000 Stuttgart 1, West Germany

For economic mass production by a microbial process, e.g., of single cell protein from a petroleum fraction it is important to find a type of reactor that meets the following requirements:

- large mass-transfer rate for oxygen and substrate at low energy input;
- the reactor must be capable of being constructed and operated in large units;
- a simple and robust design, which is characterized by low construction costs, easy to keep sterile, low maintenance costs, and a high on-stream availability.

The loop reactor, especially the jet loop reactor, is examined with regard to its suitability as a bioreactor; in this context the O_2 transfer, the substrate transfer, the necessary energy input and the flow pattern are dealt with. The optimal operating conditions (minimum energy input) regarding O_2 conversion and jet power input under consideration of a favorable flow pattern are presented. It is shown that the controllability of the reactor meets the requirements of the culture.

As an example of economically optimal SCP production, the cultivation of yeast on n-paraffin substrate is analysed.

1 Introduction

A fermenter is a reactor for the controlled performance of a microbial metabolic reaction. Since different metabolic reactions require different conditions of cultivation it is advisable to design the reactor according to the requirements of the metabolic reaction to be carried out.

The following treatise will concentrate on the design of a bioreactor to be used for the continuous production of biomass (yeast grown on n-paraffin). It will be shown that a loop reactor is very adequate for this purpose.

The loop reactor is a well-known type of reactor in chemical process technology. Of all modifications described[1, 2], the liquid jet loop reactor is of special interest as a bioreactor.

2 Requirements for Layout and Operation

By suitable design and mode of operation of a bioreactor the technical conditions are provided so that a desirable metabolic reaction can be performed within the range given by the biological system.

In order to facilitate the generation of cell mass the following measures are necessary:

a) a nutrient solution containing certain essential components is inoculated with the strain of microorganism to be cultivated;

b) the cells must be supplied with O_2, available nitrogen, a favorable carbon source and nutrient salts;

c) the heat of reaction as well as the CO_2 generated by cell respiration must be removed, optimal environmental conditions (pH and temperature) must be maintained.

Accordingly the essential task of a bioreactor consists in the transport of mass and energy. In chemical reaction engineering it is usual to make a difference between "macrokinetics" and "microkinetics"[3]. Microkinetics usually refers to the chemical kinetics, in a strict sense, i.e., the direct interaction of molecules, whereas macrokinetics refers to the overall resulting kinetics, which includes transport phenomena like diffusion, convection, heat conduction etc. Applied to the conditions in a bioreactor, microkinetics would refer to the maximal specific growth rate, μ_{max}. This is achieved when each individual cell is supplied with everything necessary for its optimal growth according to its maximum assimilation rate. Everything limiting this growth rate, i.e., gaseous and liquid products resulting from metabolic activity or the heat created, must be taken away instantaneously. These conditions can be approximately achieved by intensive agitation and aeration at low cell density.

Macrokinetics deals as well with the possible mass and energy transfer rates within the reactor. A direct manipulation of the microkinetics can only be achieved by the choice of operation conditions through which effects can be induced in the immediate vicinity of the individual cell. For the layout of the bioreactor and the choice of the mode of operation it is advisable to examine the demands of the single cell first and then to specify the necessary measures to be taken to influence the conditions in the macro-range. The following aspects should be observed:

- the uniform distribution of cells and substrates in the reactor through adequately fixed convective streams as well as turbulent backmixing,
- the transfer of substrates into the liquid phase, through the liquid film about the cell, through the cell wall, into the interior of the cell as well as the backtransport of gaseous and liquid metabolites,
- removal of the heat of reaction,
- generation of a gas-liquid interface,
- breaking of the liquid film about the cell,
- disruption of cell agglomerates or prevention of their formation,
- the micro emulsification of low soluble substrates like n-paraffin.

This is achieved by introducing zones of high turbulence which affect the core of the liquid by creating micro-shear fields. The suitability of the bioreactor on a commercial scale depends on the fact that these unit operations are carried out with the smallest possible energy expenditure. In addition, the apparatus must run reliably and must render a high on-stream availability combined with low repair and maintenance costs. The apparatus ought to be simple and robust, cheap to manufacture, simple to sterilize and simple to keep under sterile conditions.

3 The Loop Reactor for the Performance of Metabolic Reactions

The loop reactor consists of a cylindrical, vertical vessel with a slenderness ratio of 1:5–1:10, and an inner concentric draft tube. The optimal ratio of inner to outer cylinder diameter is 0.5–0.6 according to[11, 12]. The draft tube is fixed inside the reactor in such a position, that there is a flow connection at the bottom and at the top of the reactor. Thus a directed flow through and round about the draft tube is facilitated. The internal liquid circulation within the loop is induced by means of a liquid jet in the type of bioreactor under discussion here[1, 2].

3.1 Energy Input and Energy Conversion

A prior condition for the realization of the described effects within the micro range is that power is brought into the reactor. This is achieved

a) by jet energy
b) by blowing in of air in order to meet the oxygen demand of the culture.

P_L stands for the kinetic power of the liquid jet; it is dependent upon the flow rate $\dot{V}_{L_1}$ which passes through the nozzle and its dynamic pressure p_1 which can easily be influenced by the nozzle diameter D_1 [1])

$$P_L = \dot{V}_{L_1}\, p_1 = \frac{8}{\pi^2} \rho_L \frac{\dot{V}^3_{L_1}}{D_1^4}. \tag{1}$$

The liquid jet has a double task
- the dispersion of the liquid gas mixture
- the distribution of the mixture over the entire reactor volume.

The liquid jets are injected in the direction of the liquid flow and so induce directly the circulation flow; the loss of energy by impact and friction at the draft tube or the outer wall which can hardly be avoided in stirrer agitated vessels is reduced here to a negligible amount. Losses by internal friction and diversion caused by the circulating flow can be kept low as long as the optimal reactor geometry is observed[11]). At the point where gas and liquid jet meet, i.e., for a ring nozzle at the mouth, powerful shear fields occur. These are of importance not only for the formation of fine gas bubbles and a resulting large gas/liquid-interfacial area. Moreover the occurrence of a large energy dissipation concentration in the suction region of the jet is important for the micro emulsification of droplets and for the disagglomeration of cell aggregates.

The power introduced by the gas flow to the two-phase mixture is[1]):

$$P_G = \dot{M}_G \left(R\, T_G \ln \frac{p_{G_\alpha}}{p_{G_\omega}} + \frac{w^2_{G_\alpha}}{2} \right). \tag{2}$$

The first term represents the power released by the rising gas bubbles via isothermal expansion. The second term represents the kinetic power of the entering gas flow. Generally the second term can be neglected. As long as the nozzle layout is advantageous the kinetic term evaluates to as little as 0.5% of the total power P_G in a reactor filled with liquid to a height of 30 m.

The height of the reactor is favorable for the O_2-transfer for two reasons.

1. The taller the reactor, the longer is the residence time of the gas bubbles, because the minimum distance they all are forced to travel is much longer than in a short compact reactor. In addition fine bubbles are swept along by the circulating liquid and perhaps recirculated several times. This brings about a favorable oxygen conversion. On the other hand, the high oxygen concentration of the feed gas and consequently the concentration gradient is lowered, which is a disadvantage (see [1]), Chap. 9.2).

2. The liquid head of the filled reactor leads moreover to an increased O_2 concentration in the liquid film at the gas/liquid interface. Both the longer residence time of the gas bubbles and the favorable driving concentration gradient due to the higher saturation of the film improve O_2 conversion. As a consequence the reactor can be run with a low aeration rate, which leads to saving of compression energy.

Due to its internal fluid recirculation the loop reactor has a favorable mixing efficiency[30, 41]. That is why it is not necessary – as in the case of certain types of bubble column bioreactors – to inject gas for the intermixing of the phases which of course implies an uneconomic conversion of oxygen.

3.2 Constructional Features and Suitability for Operation

The loop reactor is of a simple design and can be manufactured at low costs. All necessary installations can be mounted in vertical position; that is why it can be cleaned and steamed easily. Any piling up of material inside can be avoided, so that the apparatus can be heated up for sterilization in a short time.

No moving parts are necessary inside the vessel. There is no need for an elaborate shaft sealing – as has to be provided in the case of a stirred tank bioreactor – nor for the problematic shaft bearing within the vessel. That is why the loop bioreactor is far less susceptible to troubles than mechanically stirred reactors. It is easy to maintain sterile operating conditions and to take the necessary steps towards sterilization. The superiority over the mechanically stirred vessel is of special importance for units in the hundreds of cubic meters of liquid contents, which are needed for the bulk production of biomass. The maintenance of circulating pumps installed outside the vessel is easier. There is no need for special measures to prevent the penetration of outside microorganisms through the shaft-sealing of the pumps; a double slide ring sealing with a superimposed sterile sealing liquid circulation is sufficient. Even for very large circulation rates, reliable standard pumps are available. This affects the installation costs as well as the maintenance requirements favorably and improves the on-stream availability of the bioreactor.

Due to its slender construction the loop reactor has a favorable ratio of wall area to contents. For the removal of the heat of reaction the wall of the vessel and – if necessary the draft tube or tubes – can be used as heat exchangers (Fig. 1)[7, 8].

3.3 Controllability

For the aeration of liquids which due to their composition prohibit the coalescence of bubbles (noncoalescing systems) only the specific energy input influences the mass-transfer rate[6]. It could be proved[1, 2, 9] that the generation of the gas/liquid interfacial area for the oxygen transfer correlates especially well with the energy input P_L for a jet loop reactor.

The internal circulation rate $\dot{V}_2$ of a two phase jet loop reactor can best be related (as measured up to $w_G \approx 10$ cm s^{-1}) to the Re number Re_1 based on the nozzle diameter D_1[28, 29]. The circulation around the internal draft tube is important for the mixing of the phases and – as will be shown later on – advantageous for achieving a high productivity rate during such metabolic reactions which tend to the formation of cell aggregates, and for the utilization of substrates, which are almost insoluble in water.

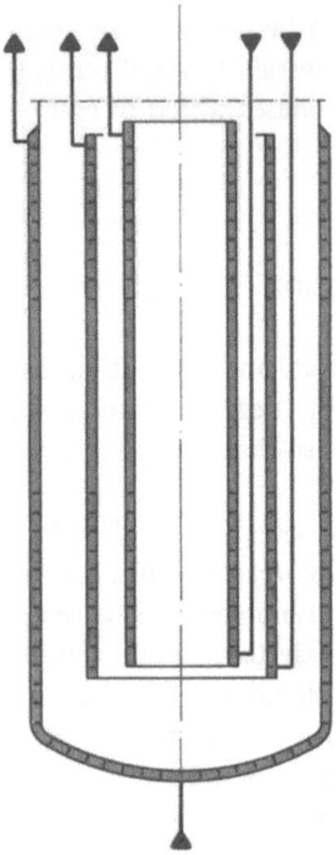

Fig. 1. Arrangement of internal heat exchangers in a loop reactor[4]. Besides the wall of the vessel also the draft tube and if necessary a second tube concentric to the draft tube can be made in the form of coolers with their own inlet- and outlet-connections each for cooling water

Both effects can be adjusted independently according to the respective demand if two-phase nozzles and pure liquid jet nozzles are combined in a certain geometric arrangement (Fig. 2)[10]:

- the two phase nozzles mainly for the primary dispersion of the gas and liquid phase,
- the liquid jet nozzles mainly for the propulsion of the circulation flow and consequently the distribution of the phases and furthermore for the redispersion of all dispersed phases.

In that way the reactor can easily be adjusted to the optimal point of operation, the energy input can be regulated in a flexible manner according to the respective demands of the culture.

3.4 Flow Pattern

A directed internal circulation flow is achieved by means of the concentric draft tube inside the reactor. The hydrodynamics are easy to understand and are predictable[1, 11, 12, 28]. By means of adequate constructional measures the formation of dead zones and the settling of solids can be avoided. Foam generation can be suppressed by strict observation of fixed operating conditions and be supported by the installation of baffles at the flow diversion round the draft tube[13].

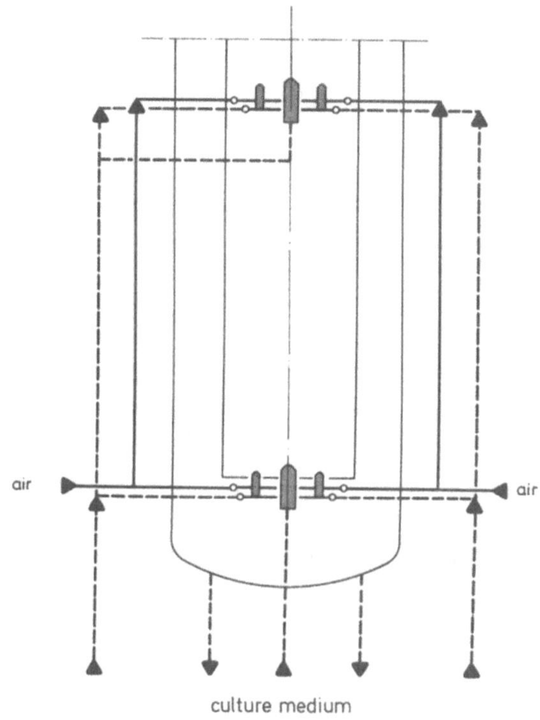

Fig. 2. Combination of pure liquid nozzles and gas/liquid nozzles in a jet loop reactor[10]. Centrically mounted pure liquid nozzles are surrounded by two phase (gas/liquid) nozzles mounted on a concentrical ring

4 Fundamentals of Design

To enable an adaptation of the construction and the operating conditions to the requirements, some data concerning the metabolic process to be carried out must be given.

4.1 Data Concerning Microbial Growth

In an unlimited system the biological data of the strain to be cultivated have to be determined; these are

- maximal specific growth rate μ_{max} and
- the cell yield Y.

The specific O_2 demand can be calculated according to[14] from the elementary analysis of cell mass and of the substrate be means of the following equation:

$$f = \frac{32\,C + 8\,H - 16\,O}{Y\,M} + 0.01\,O' - 0.377\,C' + 0.01714\,N'. \tag{3}$$

It is valid provided that besides cell mass only CO_2 and water are formed and NH_3 is used as the nitrogen source (Fig. 3). From the specific O_2 demand the heat of reaction

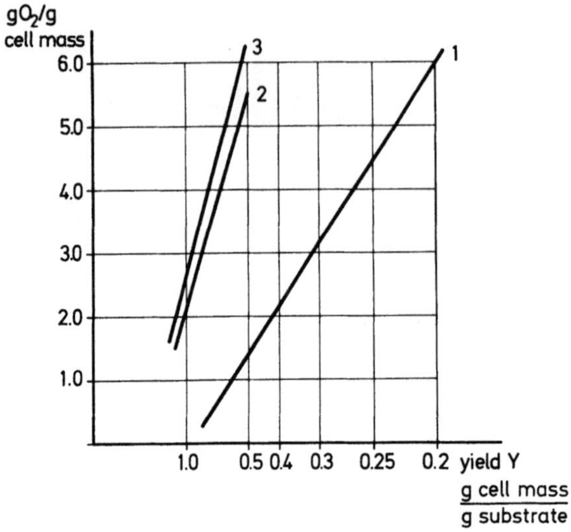

Analysis of cell mass (dry basis)

	wt % bacteria	wt % yeast
O	19	30.5
C	53	48
N	12	8.8
H	7.3	6.2
rest ash		

curve 1 bacteria grown on methanol*
curve 2 yeast grown on hexadecane
curve 3 bacteria grown on methane*

* Mateles: Biotechnolog. Bioeng. XIII 1971 S. 581

Fig. 3. Specific oxygen demand vs. cell yield for various substrates[14)]

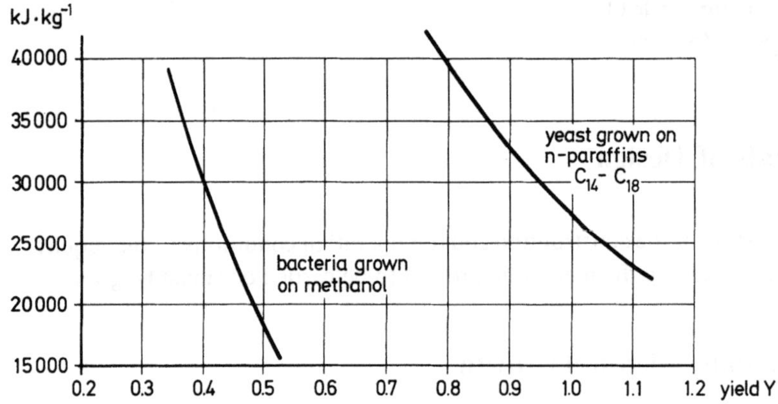

Fig. 4. Heat of reaction vs. cell yield

can be deduced (Fig. 4):

$$q = 14.38\ f \quad \mathrm{kJ\ kg^{-1}}\ . \tag{4}$$

Applying Hess' rule to microbial growth, the heat of reaction will be the difference between the heat of combustion of the substrate and of the cell mass generated[16)].

$$q = \frac{H_{o_s} - Y\,H_{o_z}}{Y}\ . \tag{5}$$

The accuracy of the data evaluated by this method is sufficient for the purpose of designing the bioreactor.

4.2 Physical Data of Media

For prediction of the gas-liquid mass transfer, of the generation of the interfacial area and of the fluid dynamics in the reactor, knowledge of the physical data of the culture medium is important.

Table 1 gives the viscosity of a yeast culture grown on n-paraffin at 30 °C at a cell concentration of $x = 20\ g\ l^{-1}$. As a comparison the viscosity of water and of a sulfite solution ($100\ g\ l^{-1}$) are quoted at the same temperature.

Table 1. Viscosity

	Viscosity ($m^2\ s^{-1}$) at 30 °C	Source
Medium, containing $20\ g\ l^{-1}$ yeast	1.11×10^{-6}	a
Distilled water	0.800×10^{-6}	17)
Sulfite solution ($100\ g\ l^{-1}$)	1.056×10^{-6}	9)

[a] Measurement by means of capillary viscosimeter according to Ubbelohde (DIN 51 562)

The surface tension of culture media containing living yeast or bacteria is given in Table 2. It contains a compilation of measured data. The samples were taken out of the reactor at different times. The respective cell concentration is quoted as well.

Table 2. Surface tension

	Yeast/bacteria	Cellconcentration of culture ($g\ l^{-1}$)	Temperature (°C)	Surface tension ($dyn\ cm^{-1}$)
Filtrate[a]	*Methylomonas sp.*	–	18	58[b]
Culture medium containing cells	*Methylomonas sp.*	12	18	57[b]
		12	37	51[b]
Filtrate[b]	*Endomycopsis lipolytica*	–	30	48[b]
Culture medium containing cells	*Endomycopsis lipolytica*	20	30	32[b]
Distilled water	–	–	30	71.2
Sulfite solution ($100\ g\ l^{-1}$)	–	–	30	60.8

[a] Culture medium, cells being removed
[b] Measured by means of a tensiometer according to Lecomte du Noüy (tearing off method)

It is obvious, that the growing cells release metabolic products into the culture medium that reduce surface tension. Surface tension is reduced more in cultures of paraffin metabolising yeast than in cultures of bacteria, grown on methanol. If the cells are removed from the culture medium in which they have grown, a larger surface tension is found than in the original culture medium. The presence of cells at the boundary layer thus reduces surface tension even further.

The density of the culture medium is shown in Table 3. The density is not very different from that of pure water, as could be expected. The cell concentration is small and because of their high water content the density of the cells is almost the same as that of nutrient solution, the density of which is in turn, – because of its low salt concentration – not very different from that of distilled water.

Table 3. Density

	Temperature (°C)	Density (kg m^{-3})
Culture solution with $x = 20\ g\ l^{-1}$ cell concentration (yeast)	30	998
Distilled water	30	995.8 [17]
Sulfite solution (100 g l^{-1})	30	1,095 [9]

Henry's constant He, the distribution coefficient between the gas and liquid phase depends on temperature and salt content of the liquid phase. Table 4 quotes He of oxygen in a yeast culture medium with a cell concentration of $x = 20\ g\ l^{-1}$. A calculation for another salt content can be carried out on the basis of an equation published by van Krevelen and Hoftyzer[18]

$$\frac{\ln c_s}{\ln c_w} = - k_s I . \tag{6}$$

k_s stands for a sum of independent items, which are characteristic of the positive and negative ions in the solution and for the solved gas.

$$k_s = i_+ + i_- + i_g . \tag{7}$$

The total ionic strength is defined by

$$I = \frac{1}{2} \Sigma c_i z_i^2 . \tag{8}$$

Table 4. Henry's constant for O_2 in different solutions

	Temperature (°C)	
Filtrate from 20 g l^{-1} cell mass containing yeast culture	30	40.0[a] [19]
Distilled water	30	39.3
Sulfite solution (100 g l^{-1}	30	69.0 [9]

[a] O_2 concentration measured according to Winkler[20]

The diffusion coefficient D_L for molecular diffusion of O_2 in a nutrient medium was determined experimentally. The measurements were carried out in a thermostabilized diaphragm cell under quasi stationary diffusion conditions[19].

Table 5. Diffusion coefficient D_L for O_2 in different salt solutions

	Temperature (°C)	Diffusion coefficient D_L ($m^2\ s^{-1}$)
Nutrient medium[a]	30	1.9×10^{-9} [19]
Distilled water	21	2.33×10^{-9}
Sulfite solution	30	2.1×10^{-9} [9]

[a] Composition see Table 6

For other compositions of nutrient salts the diffusion coefficient can be predicted according to Rattcliff and Holderoft's[21] equation:

$$\log \frac{D_W}{D_L} = B \log \frac{\nu_L}{\nu_W} \tag{9}$$

with B = 0.637.

The calculation for another temperature is possible if one applies the Stokes-Einstein equation:

$$\frac{D_L \cdot \nu_L}{T} = \text{const}\,.$$

Table 6. Composition of nutrient salt medium

H_3PO_4 (85 wt%)	2.0 ml
NH_4OH (25 wt%)	2.8 ml
$MgSO_4 \cdot 7\,H_2O$	0.9 g
K_2SO_4	1.4 g
Na_2SO_4	0.2 g
$Ca(NO_3)_2 \cdot 4\,H_2O$	0.3 g
Trace element solution	1.0 ml
Thiamine solution	1.0 ml
Tap water	1.0 l

If the characteristic data of the media are compared in Tables 1–5 with that of sulfite solution, noteworthy differences are found only in the surface tension and in Henry's constant He.

interface equilibrium is reached between the O_2 content of the air and the corresponding saturation concentration in the film.

$$\bar{c}_{O_2} = \frac{\bar{p}_{O_2}}{He} \tag{13}$$

with

$$\bar{p}_{O_2} = y\,\bar{p}\,. \tag{14}$$

The mean liquid head is

$$\bar{p} = p_N + \frac{\gamma_L H_L}{2 \times 10^4}\,. \tag{15}$$

The reactor is run at a large internal circulation rate; this is necessary – as will be shown later on – because of the mixing of the phases and for the intensificiation of the substrate transfer.

That is why many fine gas bubbles are recirculated repeatedly round about the draft tube. An estimation of the bubble volume recirculated indicates that the gas phase can be regarded approximately as being totally mixed; therefore we can write

$$y \approx y_\omega \tag{16}$$

and

$$y_\omega = y_a \frac{1 - u_{O_2}}{1 - y_a} = 0.21\, f(u_{O_2})\,. \tag{17}$$

By the insertion of Eqs. (13)–(17) in Eq. (12) it follows that

$$a_L = \frac{\dot{M}_{O_2}/V_L}{3600\, k_L \left[0.21 \frac{f(u_{O_2})}{He}\left(p_N + \frac{\gamma_L H_L}{2 \times 10^4}\right) - c_{crit}\right]}\,. \tag{18}$$

5.1.1 The Oxygen Transfer Coefficient k_L

As basis for the following considerations the mass transfer coefficient k_L is regarded as constant. According to Blenke and Hirner[2], Calderbank[24], and Reith[25] k_L is independent of the hydrodynamics and of the energy dissipation within a certain range of highly turbulent systems of technical importance. This statement is valid for aerated stirred tank reactors as well as for bubble columns and jet reactors. The experiments of the authors mentioned above were carried out by means of a chemical model reaction (oxidation of sodium sulfite in aqueous solution using copper or cobalt as a catalyst). In principle this result should be valid for bioreactors with comparable flow pattern. Moreover, it can be expected that the value of the mass transfer coefficient k_L will not be very different in a bioreactor. For according to the surface renewal theory of Danckwerts[26], $k_L \sim \sqrt{D_L}$. This statement can be regarded as proven. Comparing D_L for sulfite solution and a reaction mixture (see Table 5) no significant difference can be found. In order to affirm this consideration k_L has been measured in a falling film absorber for both systems, i.e., in culture medium containing living cells and in sulfite solution[19]. The apparatus (Fig. 6) is described in[2]. The liquid flow pattern in the falling film is quite different from that of a highly turbulent fluid, as will be met in a loop reactor.

Nevertheless it can be expected that the ratio of the mass transfer coefficients determined under the same flow conditions may be useful as transformation criteria applicable to describe mass transfer under other flow conditions. k_L has been measured in a falling film absorber using a culture of bacteria as absorbent characterized by the following data:

Table 7. Data of a culture of bacteria in a falling film apparatus, composition of nutrient medium

NH_4OH (25 wt%)	2.8 $g\,l^{-1}$
H_3PO_4 (85 wt%)	3.6 $g\,l^{-1}$
K_2SO_4	1.4 $g\,l^{-1}$
$MgSO_4 \cdot 7\,H_2O$	0.9 $g\,l^{-1}$
Na_2SO_4	0.24 $g\,l^{-1}$
$Ca(NO_3)_2 \cdot 4\,H_2O$	0.32 $g\,l^{-1}$
Trace element solution	1.0 ml
Temperature	37.0 °C
Cell concentration	16.0 $g\,l^{-1}$

From 10 measurements the mean value was

$$k_L = 1.1 \times 10^{-4}\ m\,s^{-1}\ .$$

The mean value found in the same falling film absorber for sulfite solution (100 $g\,l^{-1}$ Na_2SO_3) with 10^{-4} kmol m^{-3} Co^{2+} as a catalyst was

$$k_L = 1.24 \times 10^{-4}\ m\,s^{-1}\ .$$

This result allows us to use transfer coefficient found in the loop reactor by means of the sulfite/sulfate reaction[9] for the metabolic reaction under discussion here as well. In Fig. 7 k_L for a jet loop reactor is plotted against the superficial gas velocity w_G with the specific jet power as parameter. The mean value found was:

$$k_L = 4.6 \times 10^{-4}\ m\,s^{-1}\ .$$

5.1.2 Energy Input for Generation of Gas-Liquid Interface

The gas/liquid interface of the jet loop reactor reaches maximal values (at $P_L/V_R \gtrsim 2$ kW m^{-3}) at $w_G \lesssim 3$ cm s^{-1} (Fig. 8)[1, 2].

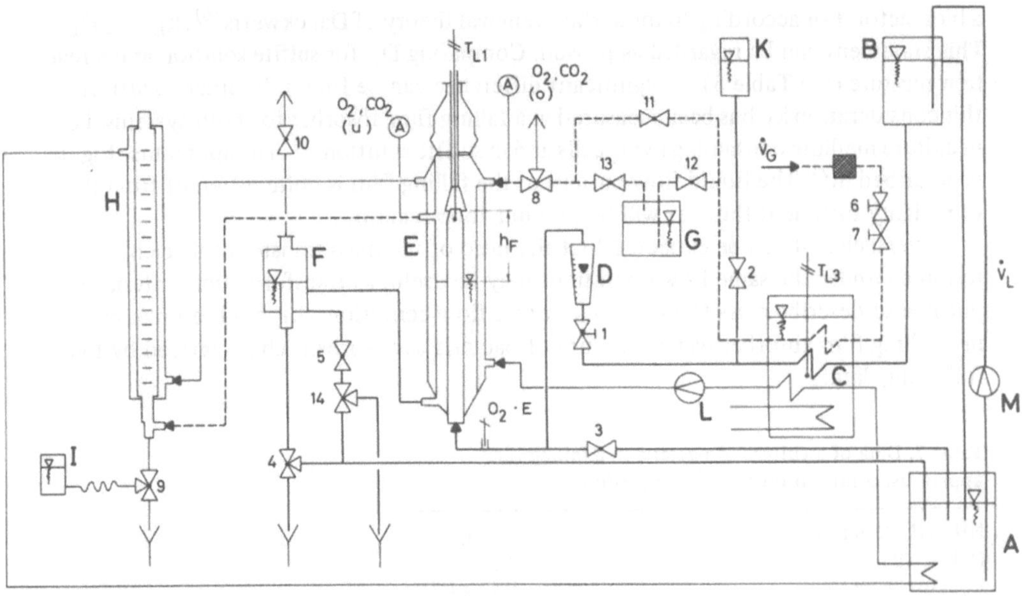

Fig. 6. Falling film apparatus for the determination of mass transfer coefficient k_L. [*A* liquid reservoir (thermo-stabilized), *B* overflow, *C* thermostat, *D* flow meter, *E* falling film absorber, *F* leveling bottle, *G* saturator, *H* bubble flow meter, *I* leveling bottle for soap-suds]

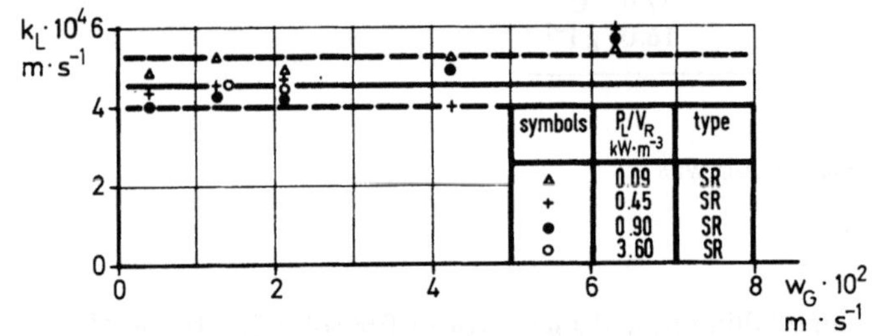

Fig. 7. Mass transfer coefficient k_L in the loop reactor with jet drive as a function of superficial gas velocity w_G and specific jet power input P_L/V_R[9]

For the sulfite/sulfate system the correlation between jet power input and interfacial area generated is in this region[8]:

$$a_R = 1500 \left(\frac{P_L}{V_R}\right)^{2/3} = 1500 \left(\frac{P_L\,(1-\epsilon)}{V_L}\right)^{2/3} . \tag{19}$$

Under these operating conditions power input by gas flow P_G does not add to the generation of interfacial area. This result is confirmed by other authors, too[6]. The application of this equation to the conditions in a bioreactor is only possible if the physical data of the fluids are taken into account.

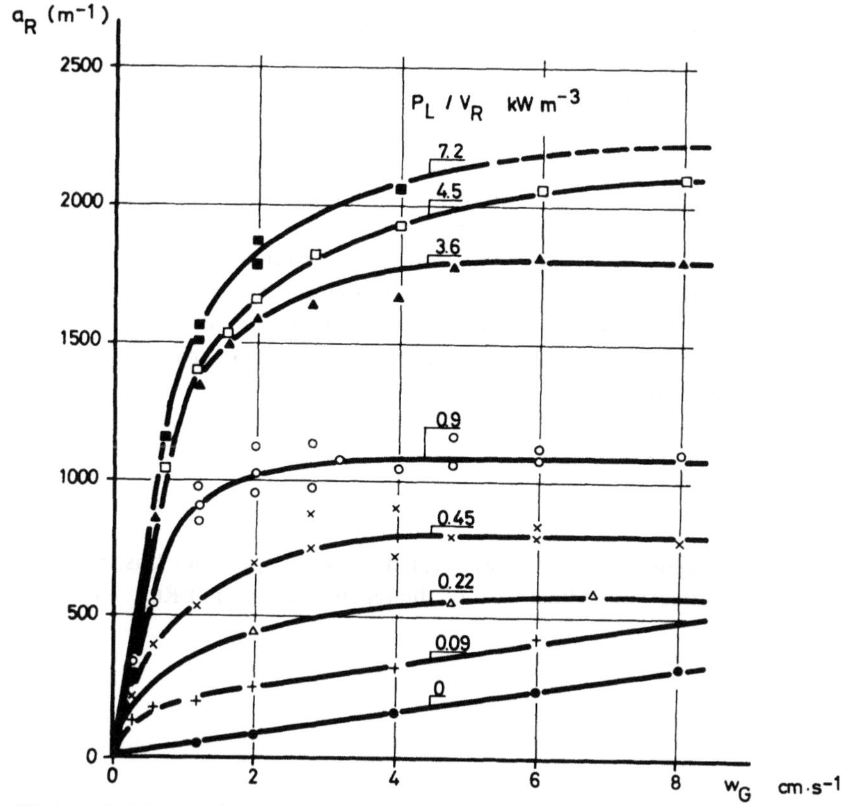

Fig. 8. Specific interfacial area of mass transfer in the loop reactor with jet drive as a function of superficial gas velocity w_G and the specific jet power input P_L/V_R[9)]

5.2 Effect of Flow Pattern on Productivity

For the generation of interfacial area necessary for the O_2 transfer the jet power input P_L is crucial, as has been shown in 5.1.2. It is important to know in which way the jet power has to be put into the reactor in order that other transfer processses may be influenced favorably. The internal liquid circulation within the loop which is important for the intermixing of the phases, correlates very well with the Re number Re_1 of the nozzle[1, 28, 29)].

Equation (1) shows that the same jet power can be introduced by means of a large volumetric flow rate $\dot{V}_{L_1}$ passing the nozzle and a low dynamic pressure p_1 or vice versa. For a given jet power a desired Re number can be adjusted by adapting the geometry of the nozzle

$$Re_1 = \frac{2}{\nu_m}\left[\frac{P_L\, D_L}{\pi\, \rho_L}\right]^{1/3} . \qquad (20)$$

If several nozzles are arranged in parallel order then

$$D_1 = (\Sigma D_{1i})^{1/2} . \qquad (21)$$

For the pilot reactor (Fig. 9) a nomogram was developed (Fig. 10) that allows us to quote the corresponding values P_L and Re_1 for a given nozzle diameter D_1.

The correlation between the internal circulation flow in an aerated loop reactor and the nozzle Re number Re_1 was determined for different gas rates $\dot{V}_G$ and different slenderness ratios s_L of the reactors[28, 29]. For characterizing the internal flow pattern within the loop a Re number of the internal liquid circulation Re_m is introduced[11], which is related to the mean flow velocity and to the mean cross section of the rising and downcoming flow.

$$Re_m = \frac{8(\dot{V}_1 + \dot{V}_2)}{\pi D_i \nu_m} . \qquad (22)$$

Figure 11 correlates $Re_m = f(Re_1, \dot{V}_G)$ at constant s_L. From this graph it can be deduced for a slenderness ratio $s_L = 7$ of the reactor operated at $Re_1 > 1.2 \times 10^5$

$$Re_1 \approx Re_m . \qquad (23)$$

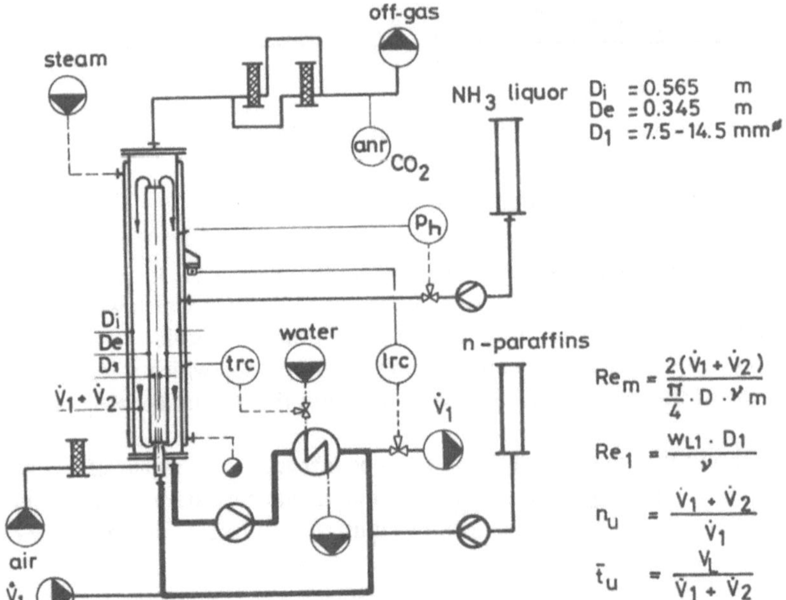

Fig. 9. Flow sheet of pilot bioreactor (jet loop reactor) (D_i inner diameter of the vessel, D_E mean diameter of draft tube, D_1 diameter of liquid jet nozzle, $\dot{V}_1$ volumetric flow rate passing the reactor, $\dot{V}_2$ liquid volume circulation flow)

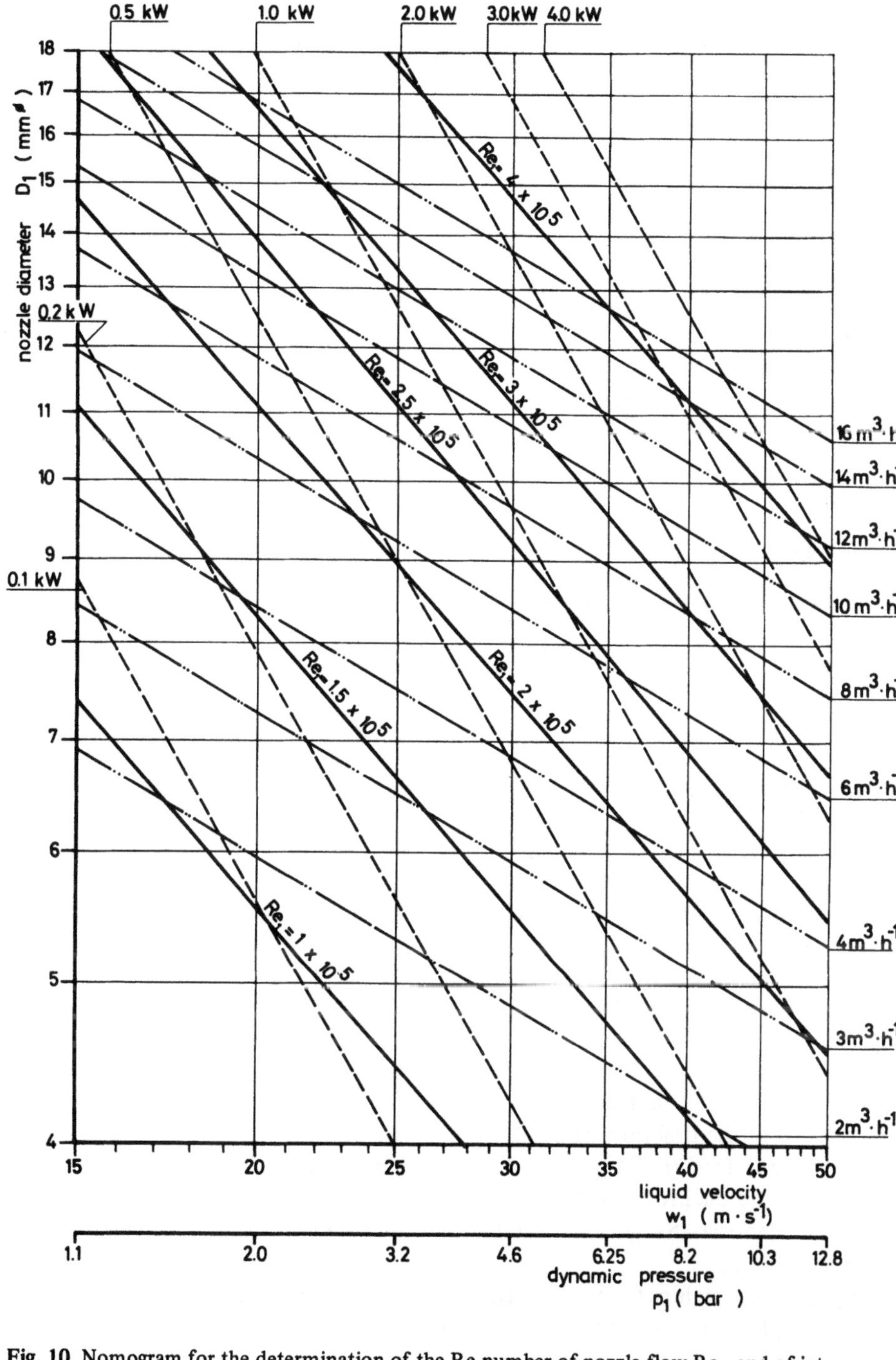

Fig. 10. Nomogram for the determination of the Re number of nozzle flow Re_1 and of jet power input P_L

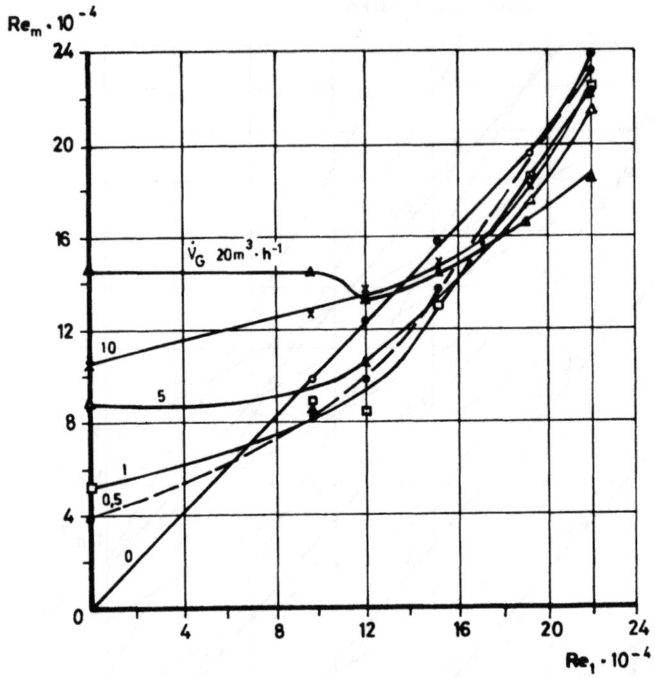

Fig. 11. Plot of liquid circulation Re number Re_m *vs.* liquid nozzle Re number Re_1 with the gas-throughput $\dot{V}_G$ as a parameter[28, 29]

For small Re numbers $Re_1 < 1.2 \times 10^5$ an influence of the gas flow rate can be recognized (airlift effect) whereas at $Re_1 > 1.2 \times 10^5$ the internal fluid circulation is mainly caused by the momentum of the liquid jet.

If this relation [Eq. (23)] is known it is possible to control the internal circulation within the loop by the choice of the nozzle diameter D_1. In a test series carried out in a pilot reactor (Fig. 9), the diameter of a centric nozzle operating in an upward direction was varied systematically and at the same time the Re number Re_1 of the nozzle flow was changed according to Eq. (20) and consequently Re_m changed according to Eq. (23). The effect of internal circulation rate on the productivity of a yeast culture grown on n-paraffin substrate was measured.

The hydrodynamic conditions in the pilot reactor are compiled in Table 8. The jet power input was at all times large enough to secure the O_2 supply of the culture.

The results of the trial runs were:

- up to a cell density of 4 g l^{-1} no dependence between the internal circulation Re number and the productivity of the culture was found,
- for cell densities $> 4 \text{ g l}^{-1}$ the productivity increased at increasing circulation Re number Re_m,
- higher jet power input at constant Re number did not lead to higher productivity.

Figure 12 plots the productivity r against Re_m [32]:

$$r = f(Re_m) . \tag{24}$$

Table 8. Hydrodynamic operating conditions of the pilot reactor

Nozzle ϕ (mm ϕ)	$\dot{V}_{L_1}$ ($m^3\ h^{-1}$)	Re_1	w_G ($10^{-2}\ m\ s^{-1}$)	w_m ($m\ s^{-1}$)	$\frac{1}{t_u}\ h^{-1}$
7.5	4.3	2.0×10^5 [a]	4.1	0.37	243
10.0	6.8	2.4×10^5 [a]	4.1	0.44	288
12.0	10.0	2.95×10^5 [a]	3.6	0.54	356
14.5	13.0	3.2×10^5 [a]	3.4	0.59	387
8.0	7.2	3.2×10^5 [b]	3.4	0.59	387

[a] $P_L = 1.26\ kW\ m^{-3}$
[b] $P_L = 2.30\ kW\ m^{-3}$

The correlation given in Fig. 12 is valid for a cell density of $x = 20\ g\ l^{-1}$ and the special geometry of the pilot reactor $H = 2.59$ m; $D_i = 0.565$ m and a concentric two-phase nozzle operating in an upward direction.

The change of internal circulation affects the flow pattern of the multiphase mixture in more than one respect and consequently the transport mechanisms in the micro range as well. The following section will examine to what kind of phenomena the observed influence on the productivity may be related. When Re_m is varied a change will occur in

a) the convective blending of the reactor contents due to a variation of the circulation frequency within the loop;
b) the turbulent backmixing as a consequence of the changed flow velocity w_m of the circulation flow;
c) the frequency of passages through the shear field surrounding the nozzle mouth.

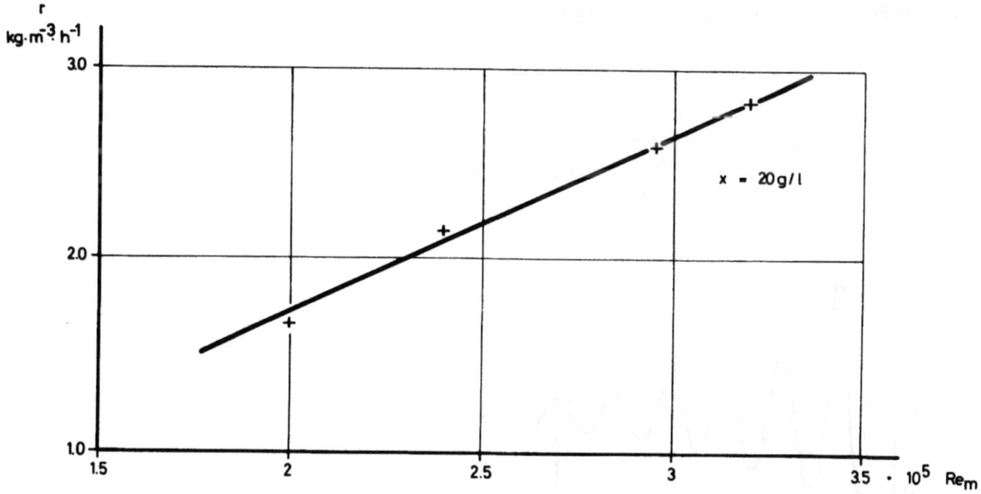

Fig. 12. Productivity of a yeast culture grown on n-paraffin vs. Re number of the circulation flow Re_m; liquid level $H_L = 2.59$ m

It is of practical interest to attain clarity on this point because a scale-up criterion can be derived therefrom.

Convective Mixing. During feed batch operation of the pilot reactor it is important to know after what time or after how many passages at a defined Re number a desired degree of mixing is achieved. According to[30, 31] the inhomogeneity h is used as standard for the degree of blending (details see[1])

$$h = \frac{\hat{s} - s_{oo}}{s_{oo}} . \tag{25}$$

The meaning of h can be derived from Fig. 13.

For the pilot reactor tested in Fig. 14 the related mixing time

$$\tau_M = \frac{t_M}{t_u} \tag{26}$$

(= number of passages i) is plotted against Re_m for h = 0.1, h = 0.01, and h = 0.001.

Herefrom it can be concluded:

a) within the range of Re numbers between 2 and 3.2×10^5 at which the reactor has been run – see shaded area Fig. 14 – a mixing efficiency of 99.9% has been reached after less than 20 passages (i.e. considering the geometry of the pilot reactor after nearly 3 min).

b) All the tests were carried out under hydrodynamic conditions, where the blending efficiency does not depend on Re_m, i.e., the reactor can be regarded as being totally mixed after a short running-in period has elapsed (cf. Table 8).

Turbulent Backmixing. In the circulating flow turbulent axial mixing is superimposed upon convective blending. According to[31]

$$1/Pe = f\,(Re_m, s_L) . \tag{27}$$

In Fig. 15 1/Pe is plotted against Re_m for different slenderness ratios s_L. It is striking that for the range $Re_m > 10^5$ $\frac{1}{Pe}$ becomes independent of Re_m and s_L.

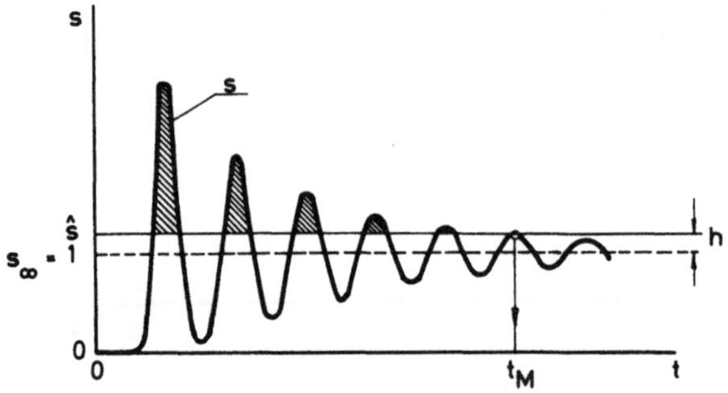

Fig. 13. Determination of inhomogeneity h characterizing the degree of mixing[31]

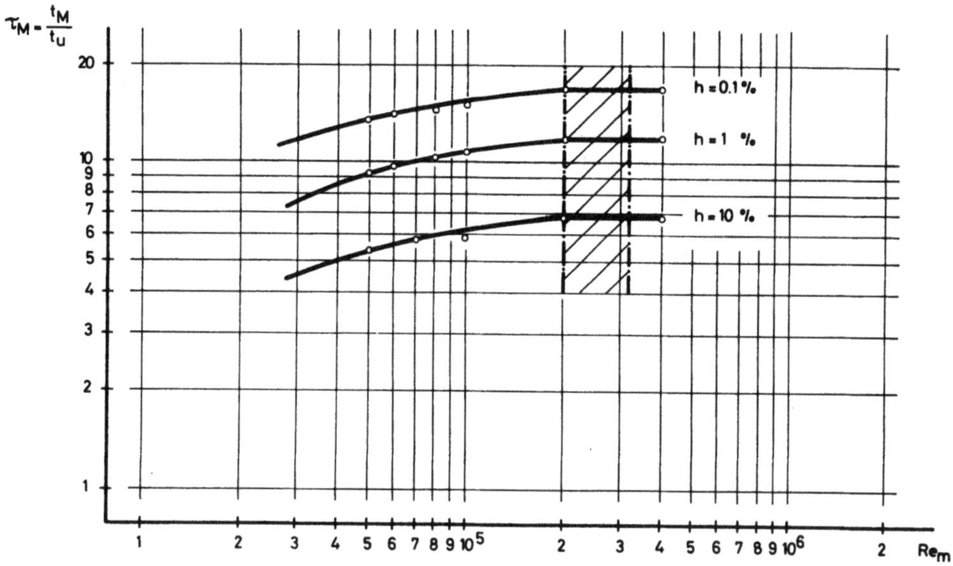

Fig. 14. Related mixing time $\tau_M = \frac{t_M}{t_u}$ in the loop reactor for various inhomogeneities

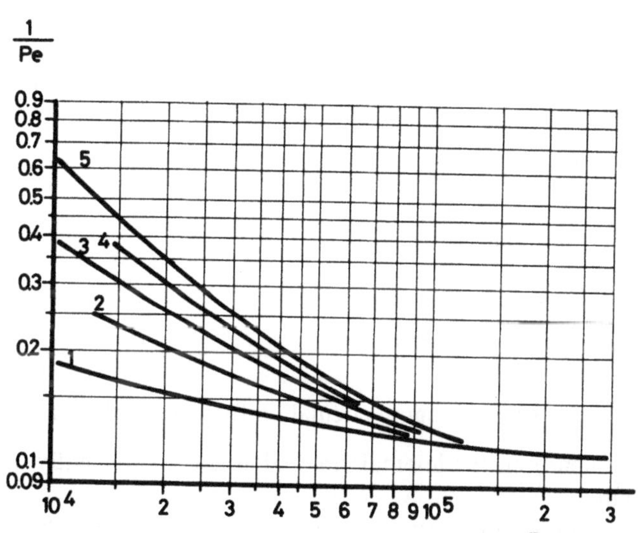

Fig. 15. Axial back mixing in the loop reactor at different slenderness ratios s_L[31] (*1* $H_L/D_i = 5$, *2* $H_L/D_i = 10$, *3* $H_L/D_i = 15$, *4* $H_L/D_i = 20$, *5* $H_L/D_i = 25$

A comparison of this result with Table 8 which contains the hydrodynamic data of the trial runs shows that the observed change of the growth rate can neither be explained by a better convective blending nor by an intensified turbulent backmixing of the reactor contents. There is good evidence for the assumption that the mass transfer mechanism is favorably affected by the frequency of passages through the shear field surrounding the liquid jet[32]. This theory that the frequency of passages is of greatest importance and not so much convective or turbulent mixing effects could be proved experimentally. This will be demonstrated later on.

If neither the flow pattern nor the degree of turbulence of internal circulation are crucial Eq. (24) should be formulated more precisely

$$r = f\left(\frac{1}{t_u}\right) = f(w_m, H_L)\,. \tag{28}$$

The frequency of passages $\frac{1}{t_u}$ is proportional to w_m and inversely proportional to the liquid height H_L.

$$\frac{1}{t_u} = \frac{w_m}{2\,H_L}\,. \tag{29}$$

In Fig. 12 r is plotted against Re_m. For the tested range of $Re_m = 2 \times 10^5 - 3.2 \times 10^5$ a proportional relationship was found. According to Eq. (28):

$$r \sim \frac{1}{t_u} \tag{30}$$

which is the same as

$$r = k\left[\frac{1}{t_u} - \left(\frac{1}{t_u}\right)_0\right]. \tag{31}$$

$\left(\frac{1}{t_u}\right)_0$ means the minimum circulation frequency which marks the beginning of liquid jet influence on the internal circulation and consequently on the productivity of the culture.

Equation (31) is based on the assumption (supported by the consideration quoted above) that the frequency of passages through the shear field surrounding the nozzle mouth plays the decisive role in the productivity of the culture.

The validity of this assumption was checked by cultivating the same yeast strain under the same conditions as the yeast cultures in Fig. 12 at a changed liquid level of the reactor $H_L \approx 2\,H_{L_1}$.

The result of these trial runs is presented as curve *B* in Fig. 16, where the productivity is plotted against the circulation velocity w_m for the technically important range of

$r > 1\ kg\,m^{-3}\,h^{-1}$. Curve *A* is valid for the liquid level $H_L = H_{L_1}$. The finding is in agreement with the above mentioned assumption: productivity of the culture in the case of the prolonged draft tube is significantly smaller than the comparable results with the short draft tube at the same circulation velocity w_m.

This finding could be confirmed by another series of trial runs: at a liquid level of $H_L \approx 2\ H_{L_1}$ an additional pure liquid jet nozzle was installed at a vertical distance of $\Delta H \approx H_1$. Both nozzles were operated at the same Re number, whereas the jet power was subdivided to one half to each nozzle. By this arrangement every particle passes twice the active region of a nozzle per circulation. The productivity of the culture found in these trial runs matches very well with curve *A*, representing the productivity attained with the short guide tube $H_L = H_{L_1}$.

Regarding these results the following conclusion is possible: If passing a nozzle has a favorable influence on the growth kinetics, then it is obviously important that as many particles as possible get seized by the active region of a nozzle per circulation. Therefore we tested a nozzle arrangement according to Fig. 20 where a large part of the cross section of the draft tube is covered by the jet cones of the nozzles.

The productivities found with these trial runs match very well curve *A* of Fig. 16. The highest productivities were attained using this nozzle arrangement.

In Fig. 17 the results from Fig. 16 are plotted vs. $\frac{1}{t_u}$. Curves *A* and *B* coincide as predicted by Eq. (31).

Table 9. Survey of nozzle arrangements tested

Curve in Fig. 16	Nozzle arrangement	Reactor geometry	Comment
A	Single centrically mounted two phase nozzle at the reactor bottom	$H_L = H_{L_1} = 2.59$ m	Results in Fig. 12 or curve *A* of Fig. 16 resp.
	Two centrically mounted nozzles; a two-phase-nozzle at the bottom and a pure liquid jet nozzle at a vertical distance of $\Delta H = 3.32$ m	$H_L \approx 2\ H_{L_1} = 5.32$ m $\Delta H_L = 3.32$ m	
	Nozzles arranged according to Fig. 20	$H_L = H_{L_1} = 2.59$ m	
B	Single centrically mounted two phase nozzle at the bottom	$H_L \approx 2\ H_{L_1} = 5.32$ m	

Similar observations concerning stirred tank bioreactors have been reported by other authors.

Chakravatry et al. for instance found a correlation between the specific growth rate and stirrer speed[33]. Einsele, Blanch, and Fiechter[34] stated in a series of comparative

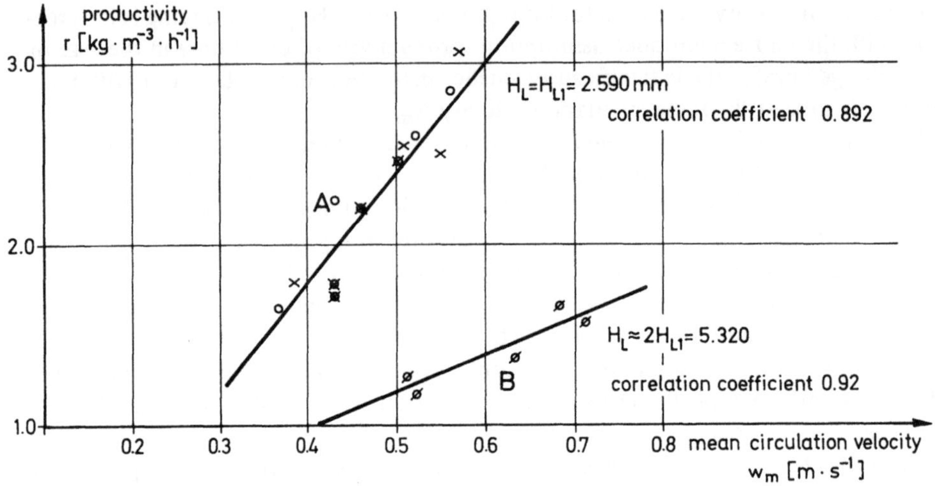

Fig. 16. Productivity of a yeast culture grown on n-paraffin vs. mean velocity of circulation flow w_m at different liquid levels and nozzle arrangements

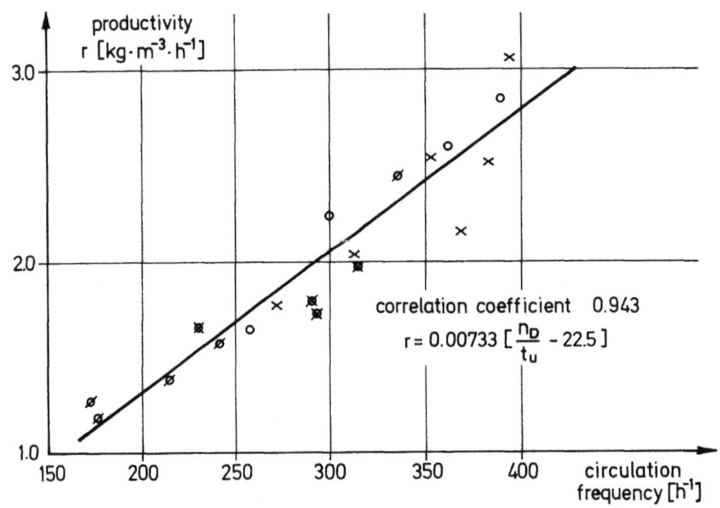

Fig. 17. Productivity of a yeast culture grown on n-paraffin vs. circulation frequency $\frac{1}{t_u}$ at different liquid levels and nozzle arrangements

experiments using *Candida tropicalis* on hexadecane using flat blade-turbines and circulation stirrers that the duration of the logarithmic growth phase in batch tests was prolonged and thus the maximum growth rate $\frac{dx}{dt}$ was positively affected by the mixing intensity.

Since the correlation between productivity and circulation frequency $\frac{1}{t_u}$ can be expressed by Eq. (31) the diameter of the liquid nozzle can be predicted according to Eqs. (29), (23), and (20). The necessary jet power P_L is then introduced in such a way that the circulation frequency as well as the formation of interfacial area for O_2 transfer are in accordance with the productivity aimed at. With a given nozzle diameter D_1 the liquid flow to the nozzle $\dot{V}_{L_1}$, that is the same as the circulation rate of the pump is

$$\dot{V}_{L_1} = 1/2 \sqrt[3]{\frac{\pi^2 P_L D_1^4}{\rho_L}} . \tag{33}$$

6 Calculation of the Optimal Operating Conditions

(Minimum Energy Input)

For economical reasons the reactor must be run as close to the optimum operating point as possible. At the same time the oxygen supply as well as the transfer of paraffin must be ensured.

6.1 Oxygen Conversion and Jet Power-Input

When considering the O_2 supply of a jet loop bioreactor it must be established whether it is more advantageous to achieve an extensive O_2 conversion by introduction of much jet power or to make use of a favorable concentration gradient between the O_2 in the gas and the dissolved oxygen in the broth by surplus of air, thus saving jet power.

The visual observation of the flow pattern in a loop reactor model with jet drive made of lucite shows that a jet of a given power can disperse optimally only a limited gas volume. If too much air is injected relative to the jet power it will not be seized by the jet and passes the shear field hardly dispersed in big bubbles. This observation was further examined in a series of trial runs using the test system sulfite/sulfate solution. In Fig. 18 the interfacial area generated relative to the total introduced power ($P_L + P_G$), is plotted versus the jet power P_L per aeration rate[1, 35, 36]. The curve exhibits a maximum at

$$P_L/\dot{V}_G = 0.01 \text{ kWh m}^{-3} . \tag{34}$$

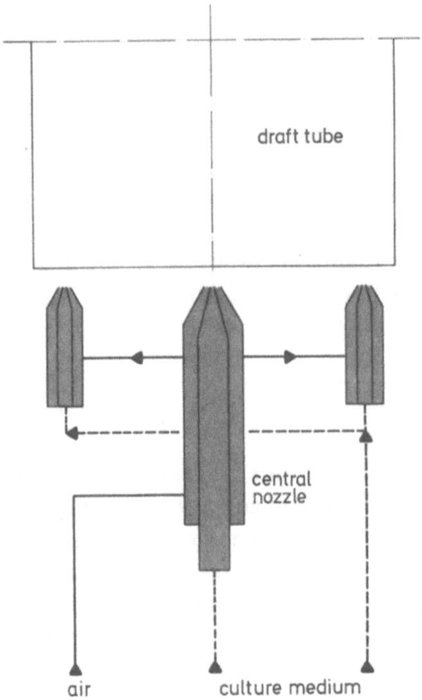

Fig. 20. Combination of a central gas/liquid nozzle surrounded by three gas/liquid nozzles symmetrically (120°) mounted on a concentric ring

b) above 30 m liquid height no further significant improvement of O_2 conversion is possible.

Inserting the u_{opt} thus evaluated in Eq. (38) yields the corresponding jet power input P_L/V_L for the transfer of a given amount of oxygen $\dot{M}_{O_2}/V_L$, and Eq. (37) gives the corresponding minimum gas flow rate. By inserting the result in Eq. (2) P_G/V_L the specific aeration power is obtained.

6.2 Economically Optimal Productivity

The productivity r found for an unlimited system cannot generally be suitably applied on a commercial scale. The criteria of importance in industrial practice are quite different from those that play a role in determing data of microbial growth on a laboratory scale. In principle on a commercial scale too a high productivity is very desirable, but the necessary expenses must be proportional to the success obtained.

In this context one has to examine the question whether it is more advantageous to aim at a large productivity by working at a large energy input and so to save reactor volume or to run the microbial process at a moderate productivity allowing for a large reactor volume and a larger harvesting capacity with energy savings. The productivity of a technical production unit ought to be based upon a calculation of its maximum profits.

As an example the economically advantageous productivity for a bioreactor with biomass production of 1500 kg h^{-1} and a specific O_2-demand $f = 2.1$ is calculated.

The calculation was based on the mean investment capital costs and energy prices valid for the German Federal Republic in 1976. The task is consequently

$$B(r) + bI(r) = \text{minimum} . \tag{40}$$

The proportional costs B(r) which are related to the productivity are more or less the energy costs B_e.

They consist of

- the power consumption for the air compressor for the O_2 supply B_{e_g}.
- The power consumption for the circulation pump providing the jet energy B_{e_L}.
- The power consumption for the harvesting equipment.

The investment I(r) as far as it is dependent on the productivity consists of

- the investment for the reactor $I_f(r) = f(V_L)$;
- the proportionate investment for the facilities for cell harvesting and cell mass concentration $I_a(r)$.

The energies are evaluated for the optimal operating point of the reactor demonstrated in 6.1. A certain necessary O_2 transfer rate is required for each productivity

$$r f = \dot{M}_{O_2}/V_L . \tag{41}$$

For optimal O_2 conversion, i.e., $u_{O_2} = u_{O_2 opt}$ there is a necessary amount of air $\dot{V}_G$ which must be blown into the reactor and which follows from Eq. (37). If one knows $\dot{V}_G$ and the liquid head $p = p_N \dfrac{H_L \gamma_L}{10^4}$ the aeration power P_G can be calculated and if one considers the efficiency factors for the compressor and its electric drive the terminal power B_{e_g} can be given.

The jet power is derived from Eq. (38). If one considers the efficiency factors for the pump and its electric drive one gets the necessary terminal power for the circulation pump B_{e_L}.

The capital costs were calculated in the following manner

$$V_L = \frac{R_P}{r} \tag{42}$$

and

$$V_{R_{geom}} = 1.2 \frac{V_L}{1 - \epsilon} . \tag{43}$$

The factor 1.2 stands for the foam-gathering space at the top of the vessel. $I_f(r)$ besides the investment for the reactor of the volume $V_{R_{geom}}$ comprises the proportionate costs for the foundations and the steel structure. $I_a(r)$ stands for the investment requirements for the machinery (separators, decanters) used for the mechanical concentration of cell mass and includes the proportionate costs of measuring and control, for the piping and the foundations. Figure 21 shows the specific aeration power P_G/V_L and specific jet power P_L/V_L as a function of the specific O_2 transfer rate $\dot{M}_{O_2}/V_L$.

In Fig. 22 the annual costs in 1000 DM p.a. for the example mentioned above are plotted against the productivity r. The following tendencies can be derived from the calculation model:

a) within the productivity range examined ($r = 1-3\ kg\ m^{-3}\ h^{-1}$) the energy costs far exceed the capital costs. There is no minimum in that range.

b) the annual energy costs do not depend greatly on productivity. This can be explained in the following way:

- the aeration power is the greatest portion of the total power input (> 80%).
- the specific aeration power P_G/V_L is practically not dependent on productivity. If the reactor is enlarged in geometrical similarity for a given productivity each pro-

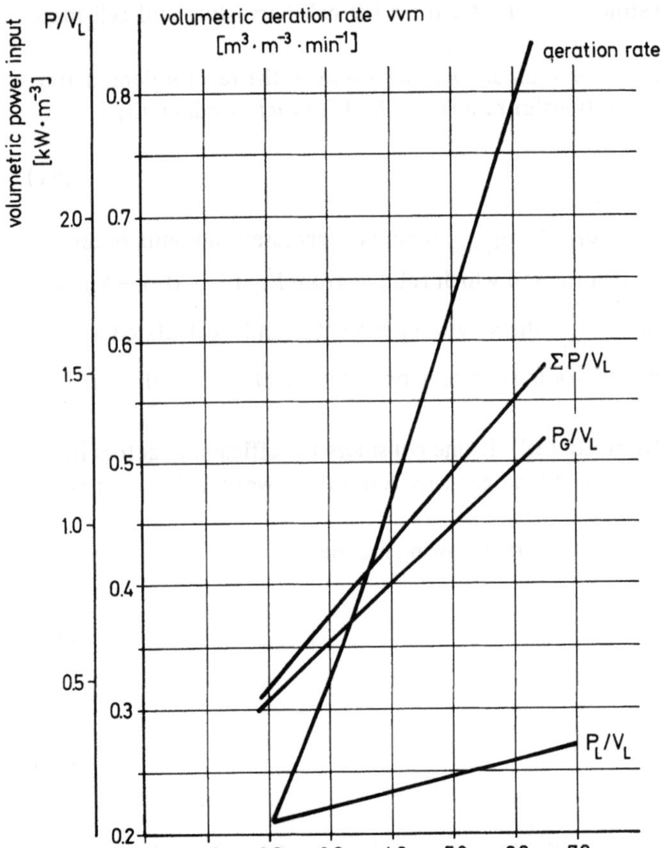

Fig. 21. Specific jet power and specific aeration power vs. specific O_2 transfer rate $\dot{M}_{O_2}/V_R$

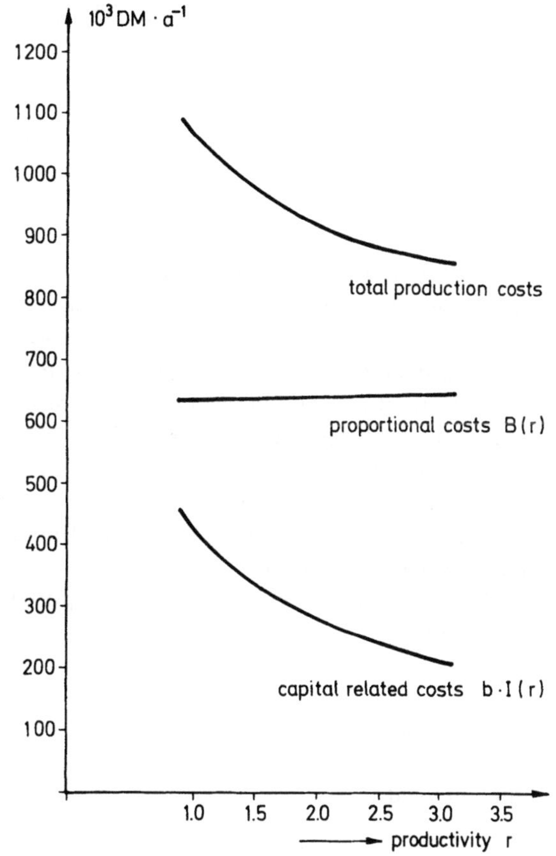

Fig. 22. Annual operating costs of a bioreactor operated at a production rate of 1500 kg h^{-1} biomass vs. productivity r

duction rate is related to a certain filling height of the reactor. The increased energy expenditure needed to allow for higher liquid head is compensated for by an increased O_2 conversion.

c) In the range r = 2.4 - 3.0 kg m^{-3} the total operating costs (i.e., Σ energy plus capital costs) depend only to a small extent on the productivity.

7 Continuous Operation in a Jet Loop Bioreactor

Because of the favorable space-time utilization of the reactor volume and the constant use made of the machinery and the capacity of the apparatus the continuous cultivation for the mass production of microorganisms is advantageous compared to batch operation. In the steady-state a constant cell density is achieved in the reactor

$$\bar{x} = Y(s_0 - \bar{s}) . \tag{44}$$

The productivity of continuous cell growth is expressed by the well known equation

$$r = D\,Y(s_0 - \bar{s}) . \tag{45}$$

By choosing D and s_0 the productivity of a continuous culture in the steady state is fixed. However, the productivity cannot be enlarged at will by increase of s_0. A speciality of cultivating yeast on n-paraffin is that after exceeding a critical cell density growth does not continue to rise exponentially in spite of the intrinsic high substrate concentration in the reactor but it follows a zero-order reaction, i.e., linearly. This phenomenon is due to a growth-limiting mass-transfer. The paraffin which is present in the reactor with a concentration $\bar{s}$ is only partly present in a form available to the cells (micro-emulsion, accomodated phase)[37, 38].

The velocity of transformation of the paraffin into an utilizable form restricts the attainable productivity of the culture.

To characterize a continuous culture it is important to distinguish between the yield of the reactor

$$Y_F = \frac{x}{s_0} \tag{46}$$

and the cell yield

$$Y = \frac{x}{s_0 - \bar{s}} . \tag{47}$$

Both are connected by the substrate conversion rate

$$u_s = \frac{s_0 - \bar{s}}{s_0} , \tag{48}$$

$$Y_F = u_s\,Y . \tag{49}$$

Reactor yield is always smaller than cell yield because substrate concentration in the reactor never reaches zero. In view of a high substrate conversion the mean substrate concentration $\bar{s}$ in the reactor should be as small as possible so that the growth rate is still not yet dependent on the concentration. This concentration was measured for the growth of yeast on n-paraffin, and was found to be roughly 500–1,000 ppm[39, 40].

7.1 Residence Time Distribution in a Loop Reactor

A very important characteristic feature of a reactor in continuous operation is its residence time behavior. The time taken for the individual components of a volume element to leave the reactor depends on the one hand on the degree by which the reactor contents are displaced by the feed $\dot{V}_1$ and on the other hand on the degree to which they are back mixed. In[1, 41, 42] an analysis of the residence-time behavior of a loop reactor is given. Based on the mass balance of the i-th passage

$$s_i(\dot{V}_1 + \dot{V}_2) = s_{i-1}\,\dot{V}_2 + s_0\,\dot{V}_1 \tag{50}$$

by introducing n_u one gets

$$s_i = \frac{1}{n_u}\,[(n_u - 1)\,s_{i-1} + s_0]\,. \tag{51}$$

For a large circulation rate $n_u \rightarrow \infty$ Eq. (51) gives

$$\lim\left(\frac{s}{s_0}\right)_{n_u \rightarrow \infty} = 1 - e^{-t/\bar{t}}\,. \tag{52}$$

The mean residence time of a volume element in the reactor is

$$\bar{t} = \frac{V_L}{\dot{V}_1} = \frac{1}{D}\,. \tag{53}$$

Equation (52) is the well-known cumulative curve of the residence time distribution of an ideally mixed stirred tank. For the loop reactor in continuous operation and without substrate consumption by growing cells one gets

$$\frac{s}{s_0} = 1 - e^{-tD}\,. \tag{54}$$

7.2 Operating Conditions for a Continuous Steady-State Cultivation

A very important operating parameter of the loop reactor for continuous cultivation is the circulation number n_u[1, 11]:

$$n_u = \frac{\dot{V}_1 + \dot{V}_2}{\dot{V}_1} = \frac{1}{t_u\,D}\,. \tag{55}$$

If by experiment the relationship between the circulation frequency $\frac{1}{t_u}$ and the productivity r of a culture which can be thereby attained are known for a given reactor geometry and arrangement of the nozzles [see Eq. (31)], then the resulting mean cell density $\overline{x}$ for the steady state in the reactor can be given

$$\overline{x} = K\,(n_u - n_{u_0})\,. \tag{56}$$

n_{u_0} here is – analogous to $\left(\frac{1}{t_u}\right)_0$ – the minimum circulation number, at which the liquid jet stream effect begins to be efficient for the internal circulation.

8 Concluding Remarks

For the economic manufacture of a mass product by a microbial process as for example single-cell-protein it is important to choose a reactor that meets the following requirements:

- large mass-transfer rate for oxygen and substrate at low energy input,
- the reactor must be capable of being constructed and operated in large units,
- a simple and robust design which is characterized by low construction costs, which is easy to keep under sterile conditions, has low maintenance costs, and a high on stream availability.

The loop reactor with jet drive meets these requirements. Before it is possible to put forward a safe design for a commercial scale bioreactor the degree of reliability of some scale-up criteria must be checked in a pilot plant and the operational suitability of certain elements must be tested over a longer period.

The energetic optimization of this type of bioreactor must still be improved upon by refinement of the geometry and the arrangement of the nozzles inside the reactor, by optimization of the D_E/D_i-ratio the specific energy demand can be diminished further.

The development of this type of bioreactor is still in its infancy; its suitability in principle has already been proved and the path future developments may take has been indicated.

Acknowledgements

This work is part of a project, jointly carried out by Hoechst AG, Gelsenberg AG (VEBA), Uhde GmbH working together on the "Microbial Production of Proteins" – research program promoted by the Federal Ministry of Research and Technology.

9 Nomenclature

A	m^2	gas/liquid interfacial area
a_L	m^{-1}	specific gas/liquid interfacial area ref. to liquid volume
a_R	m^{-1}	specific gas/liquid interfacial area ref. to reactor volume
a_s	m^{-1}	specific surface area of suspended paraffin droplets
b	%	fixed costs, related to capital investment
B	–	constant in Eq. (9)
$B(r)$	$DM\ h^{-1}$	proportional costs depending on productivity
B_e	$DM\ h^{-1}$	energy costs
B_{e_g}	$DM\ h^{-1}$	energy costs for air compression
B_{e_L}	$DM\ h^{-1}$	energy costs for providing jet power
C, H, O	–	number of atoms of carbon, hydrogen, oxygen, in the molecule of the substrate
C′, H′ O′, N′	wt %	weight percent of carbon, hydrogen, oxygen, nitrogen in the biomass
c	$kg\ m^{-3}$	concentration
c_s	$kg\ m^{-3}$	solubility of gas in an aqueous salt solution (saturation concentration)
c_W	$kg\ m^{-3}$	solubility of gas in distilled water (saturation concentration)
$\bar{c}_{O_2}$	$kg\ m^{-3}$	mean oxygen concentration allover gas/liquid interphase in the reactor
$c_{O_2}{}^*$	$kg\ m^{-3}$	liquid side oxygen concentration at gas/liquid interphase
D_1	m	diameter of liquid jet nozzle
D_L	$m^2\ s^{-1}$	diffusivity of oxygen molecules in culture medium
D_W	$m^2\ s^{-1}$	diffusivity of oxygen in distilled water
D_E	m	mean diameter of draft tube
D_i	m	inner diameter of reactor
D	h^{-1}	dilution rate $\dot{V}_1/V_L$
f	–	specific oxygen demand
F	m^2	cross section area of reactor $\frac{D_i^2\,\pi}{4}$
H_{os}, H_{oz}	$kJ\ kg^{-1}$	heat of combustion of substrate, of biomass respectively
He	–	Henry coefficient
H_L	m	liquid level of the unaerated reactor (settling height)
h	–	inhomogeneity
I	$gion\ l^{-1}$	ionic strength of solution
i_+, i_-, i_g	–	contributions of positive ion, negative ion, and gas to solubility factor k_s in Eq. (7)[18]
i	–	number of completed circulations
$I(r)$	DM	capital investment as function of productivity
$I_f(r)$	DM	capital investment for the reactor
$I_a(r)$	DM	capital investment for the harvesting equipment
k	–	constant in Eq. (6)
k_L	$m\ s^{-1}$	liquid film mass transfer coefficient
k_{LS}	$m\ s^{-1}$	mass transfer coefficient, in the liquid film surrounding paraffin droplets

K_L	$m\ s^{-1}$	overall mass transfer coefficient based on liquid side
K	–	constant in Eq. (56)
K_f	–	constant in Eq. (39)
$\dot{M}_G$	$kg\ h^{-1}$	gas flow rate
M	$kg\ mol^{-1}$	molecular weight
$\dot{M}_{O_2}$	$kg\ h^{-1}$	oxygen mass flow
n_u	–	circulation number $\frac{\dot{V}_1 + \dot{V}_2}{\dot{V}_1}$
P_L	kW	liquid jet power $\dot{V}_{L_1}\ p_1$
P_G	kW	aeration power
p_1	bar	dynamic pressure (liquid nozzle)
p_{G_α}	bar	gas pressure at reactor inlet
p_{G_ω}	bar	gas pressure at reactor exit
$\bar{p}$	bar	mean liquid head in the reactor
$\bar{p}_{O_2}$	bar	mean oxygen partial pressure in the reactor
p_N	bar	barometer reading
Pe	–	Peclet number
q	$kJ\ kg^{-1}$	reaction heat evolved
r	$kg\ m^{-3}\ h^{-1}$	productivity
R	–	gas constant
Re_1	–	Reynold's number of liquid nozzle flow
Re_m	–	Reynold's number of internal circulation flow round about the draft tube
R_p	$kg\ h^{-1}$	biomass production rate
s	$kg\ m^{-3}$	substrate concentration
s_0	$kg\ m^{-3}$	substrate concentration at liquid inlet
$\bar{s}$	$kg\ m^{-3}$	mean substrate concentration in the reactor
s^*	$kg\ m^{-3}$	equilibrium substrate concentration
$\hat{s}$	$kg\ m^{-3}$	substrate concentration at impuls maximum
s_{oo}	$kg\ m^{-3}$	mean substrate concentration after complete blending
s_L	–	slenderness ratio
T	K	temperature
t_u	s, h	circulation time $\frac{V_L}{\dot{V}_1 + \dot{V}_2}$
t_M	s	mixing time (i.e., the interval passed between tracer input and its uniform distribution)
$\bar{t}$	s	mean residence time $V_L/\dot{V}_1$
u_{O_2}	–	oxygen conversion rate $\frac{y_\alpha - y_\omega}{y_\alpha}$
u_s	–	substrate conversion rate $\frac{s_0 - \bar{s}}{s_0}$
V_{L_1}	$m^3\ h^{-1}$	liquid volume per h passing the nozzle
V_L	m^3	liquid volume of the unaerated reactor
$V_{R\,geom}$	m^3	geometric reactor volume
V_R	m^3	volume of the aerated reactor

$\dot{V}_1$	$m^3\ h^{-1}$	volumetric flow rate, passing the reactor
$\dot{V}_2$	$m^3\ h^{-1}$	liquid volume circulation flow
$\dot{V}_G$	$m^3\ h^{-1}$	gas throughput, volume under operation conditions
$\dot{V}_{0G}$	$Nm^3\ h^{-1}$	gas throughput, related to T_0 = 273 K, p_0 = 1.0133 bar
w_{G_a}	$m\ s^{-1}$	gas velocity in the annular clearance of the nozzle
w_G	$m\ s^{-1}$	superficial gas velocity, related to the cross section F of the reactor
w_m	$m\ s^{-1}$	mean liquid velocity of circulation $\frac{2(\dot{V}_1 + \dot{V}_2)}{F}$
w_1	$m\ s^{-1}$	liquid velocity in liquid nozzle
x	$kg\ m^{-3}$	cell density
Y	–	cell yield
Y_F	–	reactor yield
y	–	mol fraction of gas
y_a, y_ω	–	mol fraction of feed gas, of waste gas respectively
z_i	–	ion electric charge, valency
ϵ	–	gas hold up $\frac{V_G}{V_G + V_L}$
γ_L	$kp\ m^{-3}$	spec. gravity of medium
ν_m	$m^2\ s^{-1}$	mean kinetic viscosity of medium
ν_L	$m^2\ s^{-1}$	kinetic viscosity of a salt solution
ν_W	$m^2\ s^{-1}$	kinetic viscosity of distilled water
μ	h^{-1}	specific growth rate
ρ_{O_2}	$kg\ Nm^{-3}$	density of gaseous O_2 at T_0 = 273 K, p_0 = 1.0133 bar
ρ_L	$kg\ m^{-3}$	density of medium
τ_M	–	related mixing time t_M/t_u

10 References

1. Blenke, H.: Adv. Biochem. Eng. *13*, 121 (1979)
2. Blenke, H., Hirner, W.: VDI-Berichte Nr. 218, 549 (1974)
3. Frank-Kamenetski: Diffusion and heat transfer in chemical kinetics. New York: Plenum Press 1969
4. DOS 24 36 793 v. 31. 4. 1974
5. Nagel, O., Kürten, H., Sinn, R.: Chem.-Ing.-Techn. *44*, 899 (1972)
6. Zlokarnik, M.: Adv. Biochem. Eng. *8*, 133 (1978)
7. Blenke, H.: Chem.-Ing.-Techn. *39*, 109 (1967)
8. DOS 26 03 769 v. 31. 1. 1976
9. Hirner, W.: Stoffübergang und Stoffübergangsfläche in Gas/Flüssigkeits-Strahlreaktoren. Dissert. University Stuttgart 1974
10. DOS 25 39 502 v. 5. 9. 1975
11. Blenke, H., Bohner, K., Schuster, S.: Chem.-Ing.-Techn. *37*, 289 (1965)
12. Blenke, H., Bohner, K., Hirner, W.: Chem.-Ing.-Techn. *42*, 479 (1970)
13. DOS 26 03 668 v. 31. 1. 1976
14. Mateles, R.I.: Biotech. Bioeng. *8*, 581 (1971)
15. Cooney, C., Wang, D.I.C., Mateles, R.I.: Biotech. Bioeng. *11*, 269 (1969)

16. Abbott, B.J., Clamen, A.: Biotech. Bioeng. *15*, 117 (1973)
17. VDI-Wärmeatlas, Blatt Db 1, VDI-Verlag, Düsseldorf 1974, 2. Aufl.
18. van Krevelen, D.W., Hoftyzer, P.J.: Chim. Ind. XXO., Congr. Int. Chim. Ind. 168 (1948)
19. Reule, W.: Messung des flüssigkeitsseitigen physikalischen Stoffübergangskoeffizienten für O_2 in Mikroorganismen-Kulturlösungen. Diplomarbeit Univers. Stuttgart 1976
20. Deutsche Einheitsverfahren zur Abwasser-, Wasser- und Schlammuntersuchung G 2, 3. Aufl. Weinheim: Verlag Chemie 1960
21. Rattcliff, G.A., Holderoft, J.G.: Trans. Instn. Chem. Eng. *41*, 315 (1963)
22. Blakebrough, N.: Pure. Appl. Chem. *36*, 305 (1973)
23. Levenspiel, O.: Chemical Reaction Eng., 2nd ed. New York: John Wiley Sons Inc. 1972
24. Calderbank, P.H.: Chem. Eng., Oct. CE, 209 (1967)
25. Reith, T.: Physical aspects of bubble dispersions in liquids. Dissert. TH Delft 1968
26. Danckwerts, P.V.: Gas-Liquid Reactions, Mc. Graw Hill 1970
27. Zlokarnik, H.: Chem.-Ing.-Techn. *47*, 281 (1975)
28. Blenke, H., Bohner, K.: Verfahrenstechnik *6*, 50 (1972)
29. Bohner, K.: Gasgehalt und Flüssigkeitsumwälzung im Schlaufenreaktor. Dissert. Univers. Stuttgart 1971
30. Lehnert, J.: Verfahrenstechnik *6*, 1 (1972)
31. Lehnert, J.: Berechnung von Mischvorgängen in schlanken Schlaufenapparaten. Dissert. Univers. Stuttgart 1972
32. Seipenbusch, R., Blenke, H., Birckenstaedt, J.W., Schindler, F.: 5. Intern. Ferm. Symp., Abstr. 4.09, Berlin 1976
33. Chakravatry, M., Singh, H.D., Daruah, J.N.: Biotech. Bioeng. *17*, 399 (1975)
34. Einsele, A., Blanch, H.W., Fiechter, A.: Biotech. Bioeng., Symp. Nr. 4, 455 (1973)
35. DOS 25 41 202 v. 16. 9. 1975
36. Seipenbusch, R., Birckenstaedt, J.W., Schindler, F.: 1. Symp. Mikrobielle Proteingew. Weinheim: Verlag Chemie 1975
37. Aiba, S., Haung, K.L., Moritz, V., Someya, J.: J. Ferm. Tech. *47*, 211 (1969)
38. Einsele, A., Schneider, H., Fiechter, A.: Proc. 4. Internat. Symp. Yeasts, Part I, B 8, 91, Vienna 1974
39. Birckenstaedt, J.W., Schindler, F.: 3. Sympos. Techn. Mikrobiology, p. 307, Berlin 1973
40. DOS 23 48 793 v. 28. 9. 1973
41. Stein, W.: Chem.-Ing.-Techn. *40*, 829 (1968)
42. Stein, W.: Beitrag zur Berechnung von Schlaufenreaktoren. Dissert. Univers. Stuttgart 1968

Production of Extracellular Microbial Polysaccharides

Gary W. Pace
Biospecialties, Reading, Berks., RG1 4QU, U.K.

Renton C. Righelato
Tate & Lyle Limited, Group Research and Development
Philip Lyle Memorial Research Laboratory
Reading, Berks., RG6 2BX, U.K.

Over the past twenty years a new class of microbial products, the microbial polysaccharides, has grown in industrial importance. These products can be used as alternatives to other synthetic or natural water soluble polymers or as novel polymers in thickening, suspending and gelling applications. In this article we review the technology of the production of extracellular microbial polysaccharides with particular emphasis on the characteristics and problems associated with handling highly viscous non-Newtonian culture fluids.

1 Introduction

Millions of tons of water soluble polymers are used annually as thickening, gelling, and suspending agents. Starches, cellulosics and polyacrylamides predominate, together with substantial quantities of plant gums such as locust bean and guar. Despite their costly collection and extraction, the algal polysaccharides: alginate, carageenan, and agar-agar are used extensively. These algal gums are all anionic heteropolymers which have special properties, alginate forms gels with calcium, carrageenan with potassium and agar shows an important gelling/melting hysteresis on cooling and heating. Increasing costs of collection and volatile prices of plant and algal gums have encouraged manufacturers to look more closely at industrially produced gums such as modified starches and cellulosics and the microbial polysaccharides.

In the late 1950s and early 1960s, extensive work on microbial polysaccharides was done at the Northern Regional Research Laboratories in Peoria, Illinois. Amongst the polymers which they studied was the anionic heteropolysaccharide produced by *Xanthomas campestris* which was found to be remarkably stable to heat, acids, and alkalis and to exhibit very high viscosities at low concentrations. The polysaccharide was brought on to the market ten years later by the Kelco company and more recently by others, including ourselves. The remarkable properites of xanthan gum and the success of the product have stimulated work in many laboratories seeking improved methods of production and also other microbial polysaccharides with novel properties. However, it is not yet possible to satisfactorily predict the in use behaviour of these polymers, from their covalent structure, nor in general, from simple physical measurements. Hence, matching the chemical and physical properites of polysaccharides with the needs of the users requires sophisticated market-oriented applications technology.

In this article we review the methodology of production of microbial exopolysaccharides for industrial use and high-light the problems associated with processing highly viscous culture fluids. We introduce this subject with a brief review of the range of microbial exopolysaccharides, their possible applications and the mechanisms of biosynthesis and physiology of their production. For the purposes of this review the term exopolysaccharide refers to those polysaccharides released into the culture fluid and excludes those which are found as part of the cell wall or as a tightly bound capsule.

2 Structures and Physical Properties

Classification of microbial exopolysaccharides is difficult for two reasons: (i) the physiological roles of the majority of these polymers in vivo are unknown and the subject of considerable speculation[1,2] and (ii) there are few clear relationships between covalent structure and physical properties. For the purposes of review exopolysaccharides will be grouped into neutral and anionic polymers and sub-grouped according to component sugars and linkages. The polysaccharides listed in Tables 1 and 2 are by no means exhaustive but have been taken to exemplify the main groups produced by microorganisms. For more complete lists the reader should refer to Laskin and Lechevalier[3] and Sutherland[1].

Table 1 gives some examples of neutral polysaccharides, their basic structure, properties, and claimed uses. Several of the glucans are of commercial interest: dextran, pul-

Table 1. Some neutral microbial exopolysaccharides

Trivial name	Polysaccharide repeating unit	Organism(s)	Special solution properties and (potential) uses	Ref.
Dextran	$- Glc\ 1 \overset{\alpha}{\rightarrow} 6\ Glc -$ Some $1 \rightarrow 2, 1 \rightarrow 3$, and $1 \rightarrow 4$ linkages	*Acetobacter spp.* *Leuconostoc mesenteroides*	Plastic flow; as source for making dextran derivatives for pharmaceutical uses	3, 38, 39)
Pullulan	$\rightarrow 6\ (Glc\ 1 \overset{\alpha}{\rightarrow} 4\ Glc\ 1 \overset{\alpha}{\rightarrow} 4\ Glc)\ 1 \rightarrow$	*Pullularia pullulans*	Gels on heating; O_2-impermeable films formed; readily biodegradable	12, 40)
Curdlan	$- Glc\ 1 \overset{\beta}{\rightarrow} 3\ Glc -$	*Agrobacterium*	Gels on heating or acidification; gelled foods; stable salt and acids;	4, 12)
Scleroglucan	$- Glc\ 1 \overset{\beta}{\rightarrow} 3\ Glc -$ 6 $\beta \uparrow$ Glc	*Sclerotium glucanicum*	Highly viscous and pseudoplastic; drilling muds; latex points;	5, 12)
Cellulose	$- Glc\ 1 \overset{\beta}{\rightarrow} 4\ Glc -$	*Acetobacter spp.*	Water insoluble	41)
Levan	$- Fru\ 2 \rightarrow 6\ Fru$	*Bacillus spp.* *Leuconostoc mesenteroides*	–	3)
–	Glc, Gal, Mann, Hept, Fru, Rha	*Seratia marcescens*	–	42)
–	Glc, Gal, Rha, Mann	Pseudomonas NCLB 11264	Highly viscous and pseudoplastic	27)

Table 2. Some anionic microbial exopolysaccharides

Trivial names	Polysaccharide repeating unit	Organism	Special solution properties and (potential) uses	Ref.
Alginate	– 4D – MannA 1 $\xrightarrow{\beta}$ 4D –MannA 1 –; – 4L GulA 1 $\xrightarrow{\alpha}$ 4 L – GluA	*Pseudomonas aeruginosa* *Azotobacter vinelandii*	Range of viscosity types; gels with Ca^{2+}; textile printing, food applications	9, 10, 18, 19, 25)
Xanthan	– Glc 1 $\xrightarrow{\beta}$ 4 Glc $\uparrow_{1}^{3}\alpha$ Mann 6–OAc $\uparrow_{1}^{2}\beta$ GlcA $\uparrow_{1}^{4}\beta$ Mann 4,6 · OPyruvate	*Xanthomonas campestris*	Highly viscous and pseudoplastic; gels with galactomannans; resistant to acid, alkali, and biodegradation; oil well drilling, food industry	12, 15)
Phosphomannan	→ 4 $(Mann)_5$ 1–O–P(=O)(=O)–O–	*Hansenula capsulata* *Hansenula holstii*	Thixotrophic at high contration; gels with borax	43, 44)
–	gal, SO_4^{2-}, PO_4^{3-}	*Physarum polycephalum*	–	45)
–	4 Glc 1 $\xrightarrow{\beta}$ 4 Glc 1 $\xrightarrow{\beta}$ 3 Gal 1 $\xrightarrow{\beta}$ 4 Glc 1 – $\uparrow_{1}^{6}\beta$ Glc 1 $\xrightarrow{\beta}$ 3 Glc 1 $\xrightarrow{\beta}$ 3 Glc 1 $\xrightarrow{\beta}$ 6 Glc /\\ 4 O O 6 \\ / H_3C C COOH	*Rhizobium meliloti*	–	6)

lulan, curdlan, and scleroglucan. The last two, whilst both β (1 → 3) glucans show quite different physical properties. Scleroglucan has regular branches[4] whilst curdlan is an unbranched polymer[5]. In the case of the dextrans, molecular weight and branching are both variable[1]. The heteropolymers are generally thought to consist of repeating units, the most complex of which so far determined is the octasaccharide repeating unit of the *Rhizobium meliloti* polysaccharide[6].

Uronic acids are the principal charged components of the anionic polysaccharides listed in Table 2. The degree of charge varies from alginate, which is a polyuronide of mannuronic and guluronic acids, to the weakly acidic heteropolysaccharides such as xanthan. Succinate half esters, pyruvate ketals, and the inorganic anions sulphate and phosphate are also found. The amount of some components, such as the pyruvate ketal of xanthan and the acetyl esters of alginate, are highly variable and may give the organism a fine control over the properties of its exopolysaccharide(s). The degree of pyruvalation of xanthan gum is reported to influence the physical properties[7]. It has been suggested that the acetate groups on the mannuronic acid residues of alginic acid control the epimerisation which produces the guluronic residues[8], again influencing the physical properties, as guluronic acid blocks are thought to be responsible for gelling[9].

Of the polymers shown in Table 2, xanthan and alginates are of most commercial interest. About 8000 tonnes per annum of xanthan are currently produced for use in the food industry, in oil well drilling and numerous other areas. All of the alginate currently produced is extracted from seaweeds. However, the range of alginate products obtained from *Azotobacter vinelandii* have been found to behave similarly to those obtained from algae, and bacterial production of alginate could become a reality in the near future[10]. Growing academic and commercial interest in microbial exopolysaccharides can be expected to lead to the discovery of many more such polymers and to a greater understanding of the link between structure and physical properties.

3 Biosynthesis

Most exopolysaccharides are presumed to be synthesized by a cellular mechanism identical or similar to that involved in cell wall synthesis; a few, however, are synthesized extracellularly.

Extracellular Synthesis. The dextrans are synthesized by glucosyl transferases which transfer the glucosyl portion of sucrose to the non-reducing end of the dextran molecule, releasing fructose. Levans are thought to be synthesized by an analogous mechanism. Most of the transferase activity appears to be extracellular, with some cell-bound[11]. The predominant linkage in the dextrans is α (1 → 6) though variable branching occurs at 2, 3, and 4. Molecular weight is highly variable and can be influenced by the presence of the glucan hydrolases produced by many bacteria, including those which synthesize dextrans[12].

Cellular Synthesis. Most exopolysaccharides are synthesized by particulate enzyme systems which have as their precursors the nucleoside diphosphate sugars. These systems are analogous to those involved in the synthesis of cell wall polysaccharides. The sugars of the repeating unit are sequentially added, by specific transferases to a C55 isoprenoid alcohol phosphate. The isoprenoid alcohol is thought to be responsible for transporting the oligosaccharides from their assembly location in the cytoplasm to the outside of the outer membrane. The number of sugar units added, in the synthesis of homopolymers, or the number of repeating units built up prior to export is a matter for conjecture. The mechanisms by which the oligosaccharides are added to the growing chain, by which the chain is extruded, and by which chains are terminated are unclear. There are few studies on the molecular weights of microbial exopolysaccharides. The alginate of *Azotobacter vinelandii* appears polydisperse on gel electrophoresis[13)] and molecular weights are in the range $20–800 \times 10^3$. Sutherland[14)] reported that mutants could be obtained which made a polysaccharide with increased viscosity and presumptively higher molecular weights. Native xanthan gum has a molecular weight of approximately 15×10^6 [15)].

Sutherland[1)] has suggested that the availability of the isoprenoid alcohol carrier is a determinant of the rate of exopolysaccharide synthesis. The carrier is common to cell wall polysaccharide and exopolysaccharide synthesis and lower rates of production of exopolysaccharide in growing bacteria could be due to competition of cell wall synthesis for the carrier.

Acetate and Pyruvate Substituents. Many microbial exopolysaccharides, whilst formed of regular basic repeating oligosaccharides, have, in addition, variable modifiers such as acetate esters, pyruvate ketals and succinate half esters. The role of these substituents *in vivo* and the control of the degree of substitution is mostly conjectural. The acetate groups on the mannuronic acid residues of the alginate synthesized by *Azotobacter vinelandii* have been postulated to prevent epimerisation of mannuronic acid groups to guluronate[8)]. In other cases the relationship between substitution and biological function is unknown. However, the substitutions can have fundamental effects on the physical properties of the polysaccharide. The presence of acetate esters would be expected to influence the hydrophilic nature of the polymer, possibly a significant point in determining the interaction of the exopolysaccharide with cell wall polymers or with surfaces on to which the bacterium sticks. Acetylation can also affect the ionic behaviour of the polymer. The exopolysaccharide obtained from *Xanthomonas campestris,* xanthan, at concentrations greater than 0.5% typically shows an increase in viscosity on increasing the ionic strength of the environment, for instance from 0 to 1% NaCl. However, when the acetate ester is removed from the inner mannose of the side chain, the viscosity of the deacetylated product in distilled water and salt solution changes[16)]. The degree of substitution of the pyruvate ketal on the terminal mannose of xanthan gum varies with strain and growth conditions[7)]. Xanthan with about one in three side chains pyruvylated is reported to exhibit lower viscosities in water and salt than xanthans with all side chains bearing the acid[17)]. These effects presumably come about through an influence of the charged substituents on the aggregation of individual xanthan molecules.

As the production of microbial alginate is referred to extensively below, the presumptive pathway of its biosynthesis by *Azotobacter vinelandii* and *Pseudomonas aeruginosa* is shown in Fig. 1. The first enzyme specific to the alginate pathway is probably GDP mannose dehydrogenase. Some of the alginate negative mutants of *P. aeruginosa* have been found to lack this enzyme (Piggott, N., personal communication). It is probable that polymannuronic acid is synthesized first and the subsequent C5 epimerisation of some of the mannuronate residues is due to an extracellular epimerase[18–20]. The possible role of acetate esters in determining epimerisation has been refered to[8]. In addition the ratio of mannuronic/guluronic acid residues in polymers produced by *A. vinelandii* can be varied through calcium ion concentration which influences the activity of the epimerase[18]. The molecular weights of the polymers from *A. vinelandii* are highly polydisperse[13], and the extracellular alginate lyase[18] may play a part in determining the molecular weight distribution. Although the pathway for xanthan biosynthesis has not been established it is thought to be similar to those for other microbial exopolysaccharides[21] and is probably of the form shown in Fig. 2.

4 Process Kinetics

Microbial exopolysaccharides have generally been considered as secondary metabolites (not growth related), produced when a carbohydrate source is present in excess. Neijssel

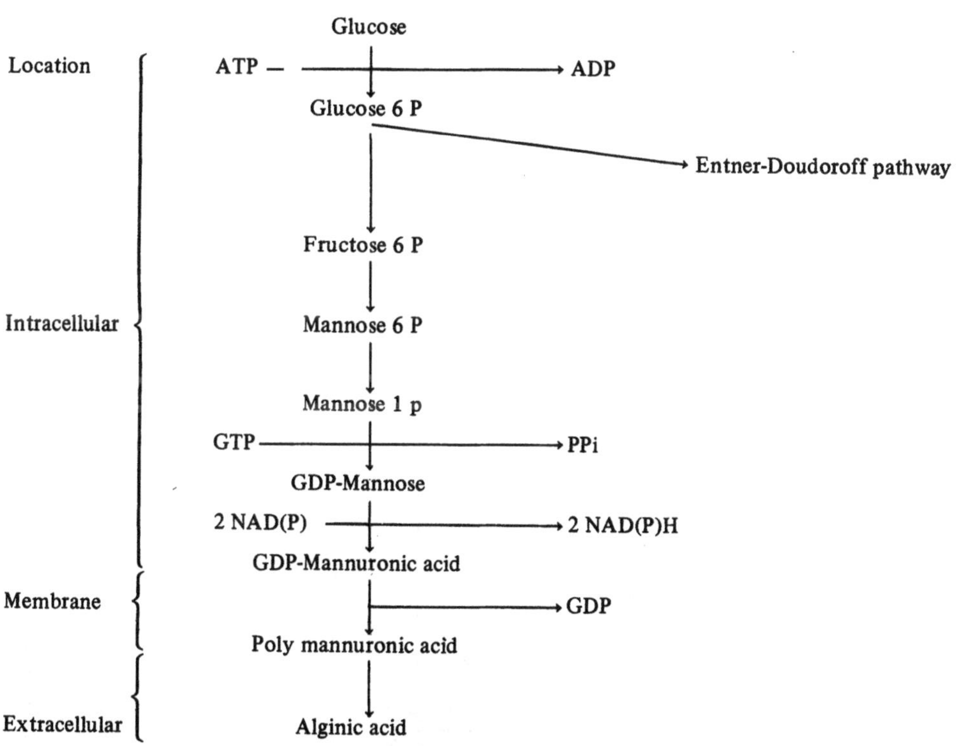

Fig. 1. Biosynthesis of alginic acid by *Azotobacter vinelandii*

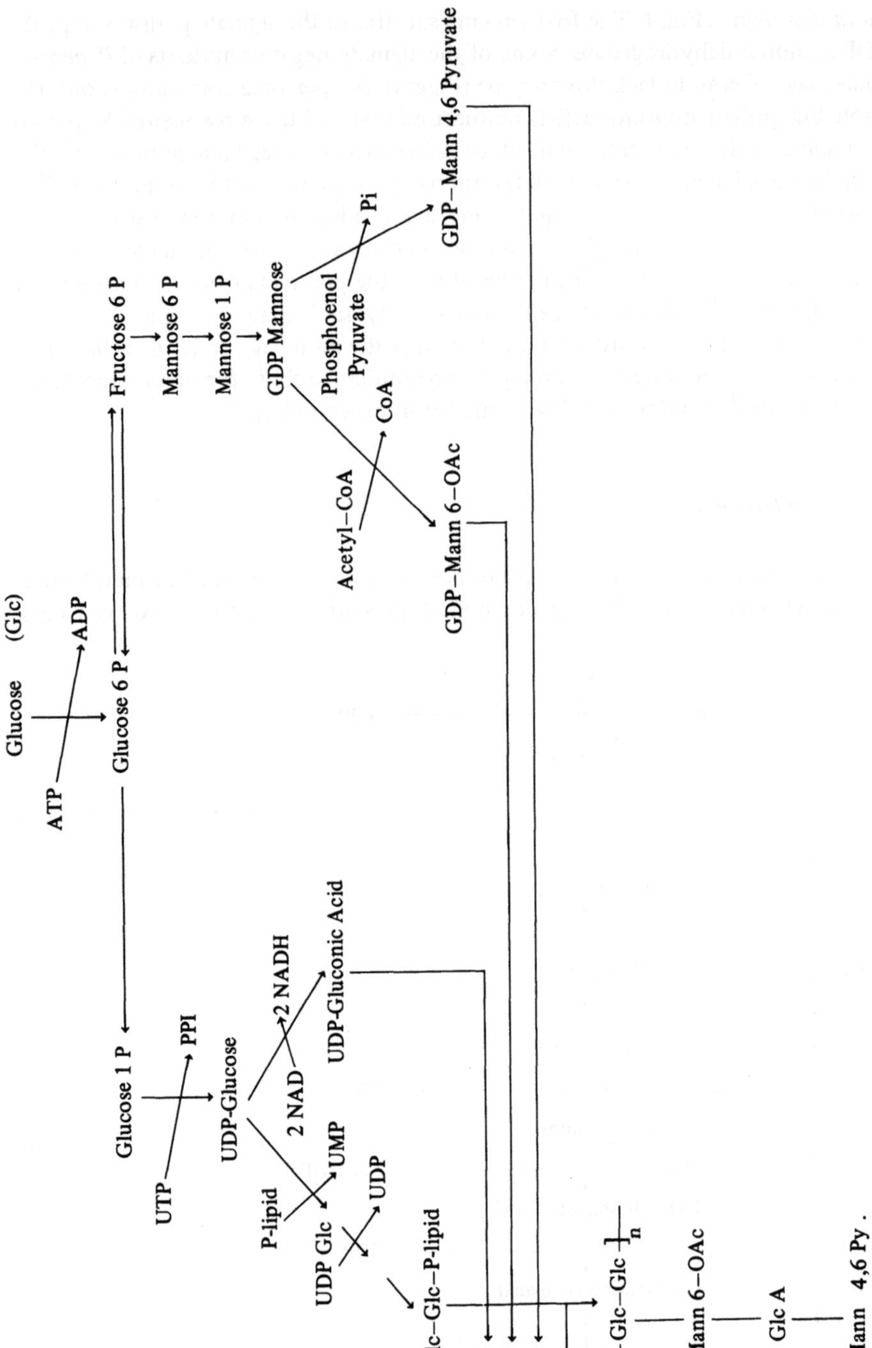

Fig. 2. Proposed pathway for biosynthesis of xanthan by *Xanthomonas campestris*

and Tempest[22] suggested from studies of *Aerobacter aerogenes* that they act as ATP sinks and are produced under conditions which would cause cells to over produce ATP, such as nitrogen-limitation of cell growth. However, recent chemostat studies of alginic acid production by *Pseudomonas aeruginosa*[23,25] and by *Azotobacter vinelandii*[2,10] and of xanthan production by *Xanthomonas campestris*[10,24] do not support these hypotheses. The results of these studies of the effect of growth rates and of growth-limiting substrate are summarized below. Studies on the effect of other environmental conditions on the kinetics of growth and product formation are also discussed below.

4.1 Growth Rate

It might be expected that a major rate-controlling step of most metabolic activities is the specific growth rate of the organism. A network of control mechanisms exist which permit the microbe to assimilate substrates, synthesize intermediates and form polymers at rates which produce more cellular material of the same type in the face of enormous environmental changes. Jarman et al.[2] showed that for *A. vinelandii* the specific rate of alginic acid synthesis was independent of specific growth rates over a range of dilution rates from 0.05–0.25 h^{-1}. Similar results were obtained by Williams[23] for a *Pseudomonas sp.* which produced an exopolysaccharide containing glucose, galactose, acetate, and pyruvate; this organism also produced polysaccharide in a non-growing system. Deavin et al.[10] reanalysed the results of the continuous culture studies of Silman and Rogovin[24] and showed that the overall rate of xanthan gum production by *X. campestris* was virtually unchanged over the growth rate range 0.05–0.20 h^{-1}. The quantity of polysaccharide produced per unit of cell mass thus increased with decreasing growth rate and was several times the cell mass at the lower dilution rates. By contrast Mian et al.[25] found that *Ps. aeruginosa* produced the same quantity of alginic acid per unit of cell mass over the growth rate range 0.05–0.10 h^{-1}; at all growth rates the ratio of polysaccharide to cell mass was about 5.

4.2 Growth-Limiting Substrate

Chemostat studies of alginic acid synthesis by *A. vinelandii*[2,10] showed that substantial amounts of exopolysaccharide were produced with N_2, MoO_4^{2-}, PO_4^{3-}, Fe^{2+}, SO_4^{2-}, Ca^{2+}, K^+ as the limiting substrates. Surprisingly it was found that in carbohydrate limited cultures polysaccharide was produced at as high a rate as with most other limitations (Table 3). Substantial amounts of alginic acid were also produced by *P. aeruginosa* under carbohydrate limited growth conditions[25] and similarly Davidson[26] reported xanthan production by *X. campestris* under carbon limited growth conditions. It appears that with these polysaccharides the organisms continue to produce them under a wide range of physiological conditions. Some of these conditions, such as carbohydrate limitations, represent a severe drain on nutrients available to the cell. By contrast Williams and Wimpenny[27] reported that no detectable polysaccharide was produced

Table 3. Exopolysaccharide production in carbohydrate-limited and excess conditions

Organism	Carbohydrate-limited		Carbohydrate excess (N-limited)	
	Cell $g\ l^{-1}$	Polysaccharide $g\ l^{-1}$	Cell $g\ l^{-1}$	Polysaccharide $g\ l^{-1}$
Azotobacter vinelandii[a]	1.3	2.2	1.5	2.0
Pseudomonas aeruginosa[b]	1.5	5.8	1.3	8.9
Xanthomonas campestris[b]	1.1	2.7	1.6	7.0

Dilution rate:
[a], $0.15\ h^{-1}$
[b], $0.05\ h^{-1}$
Synthetic media were used in each case (see text for references)

by a *Pseudomonas sp.* under glucose-limitation, whilst under ammonium-limitation 43% glucose consumed was converted to exopolysaccharide.

The type of growth-limiting substrate has been reported to influence the composition of the polysaccharides produced by *Xanthomonas sp.* Ellwood et al.[28)] observed different quantities of rhamnose in the polysaccharide(s) produced by *X. juglandis* when the organism was grown on different limiting substrates.

Davidson[26)] reported widely varying pyruvate contents of the xanthan gum produced by *X. campestris* under different growth limitations. Williams and Wimpenny[27)] did not observe significant variations in the composition of a Pseudomonas exopolysaccharide under ammonium, phosphate- and glucose-limited growth conditions.

Several reports of batch culture experimentation indicate that exopolysaccharide synthesis is maximal at temperatures below the growth rate optimum[29-31)]. In a series of chemostat experiments Williams and Wimpenny[27)] found a fairly sharp temperature optimum around 30 °C for polysaccharide synthesis by a *Pseudomonas sp.* which grew between 15 °C and 40 °C.

pH has been observed to influence polysaccharide production more than cell growth by Silman and Rogovin[32)] with *X. campestris* and by Williams and Wimpenny[27)] with *Pseudomonas sp.* pH control is likely to be needed for most production processes as the polysaccharides of commercial interest are generally acidic.

Xanthan gum is the exopolysaccharide on which most is published. It appears from the literature that batch cultivation with complex media is favoured[33, 34)]. However, simple synthetic media may be used[7, 28, 35)]. The batch growth takes 20–50 h; polysaccharide accumulation starts during the initial growth phase and continues after growth has slowed or stopped, generally giving a final xanthan gum concentration of around $20\ g\ l^{-1}$. The conversion efficiency of carbohydrate, usually glucose, to xanthan is 70–80%. The first single stage continuous culture systems to be described were genetically unstable, a poor xanthan producing organism being quickly selected[24)]. A two-stage culture, in which the first stage was carbohydrate-limited, was reported to avoid the selection of poor xanthan producing bacteria[35)], however, the published data refer to a continuous culture of short duration. More recently single-stage continuous cultures stable over many hundreds of hours have been reported[25, 28]. For a variety of reasons, including instability of industrial strains and contamination risks in production scale equipment, industrial practice is to use batch cultures or a fill-draw technique[37].

Although the microbial polysaccharides are produced industrially for their rheological properties these properties present special problems in the technology of their manufacture. In the following sections we discuss these and other problems encountered in the design and operation of production processes with particular reference to bioreactor design and product recovery.

5 Medium Rheology

Polysaccharide cultures are typified by highly viscous, non-Newtonian broths. The viscous nature of the medium is due to the presence of the polymer, which is continuous with the water phase, and particularly in the case of bacterial cultures the microbial cell contributes little to the viscosity. These fluids are distinct from fungal and streptomycetes culture medium in which the mycelium causes the broth to be highly viscous and is discontinuous with the water phase. Thus, although mycelial and polysaccharide media may show equivalent measured viscometric behaviour, the heat and mass transfer characteristics of the broth may differ markedly. For example, Blakebrough et al.[46] found that even though heat transfer coefficients of xanthan broths measured in the laminar regime were in good agreement with a theoretical prediction accounting for fluid rheology, the heat transfer coefficients for *Aspergillus niger* broths were up to four times the predicted values. Both types of broths showed similar pseudoplastic behaviour over the shear rate ranges measured.

There are few reports in the literature characterizing the changes in rheological properties which occur throughout polysaccharide production phase and furthermore, these appear confined to shear measurements of viscosity. Typically, microbial polysaccharides appear to exhibit pseudoplastic behaviour. For example, pullulan medium is pseudoplastic and fits the power law equation[47, 48] and although xanthan shows an apparent yield stress, the viscosity of the broth is also well modelled by the power law equation over a wide range of shear rates[50]. The change in viscosity over the course of a typical xanthan batch culture is shown in Fig. 3 a and 3 b. The apparent yield stress of a xanthan solution obtained by extrapolation of low shear rate measurements is typically about $2.0\,N\,m^{-2}$ for at $10\,kg\,m^{-3}$ xanthan solution in $5\,kg\,m^{-3}$ of sodium chloride. Based on qualitative observations, Charles[51] has reported that xanthan broths exhibit marked thixotropy and a Weissenberg effect, i.e., the broth climbs the impeller shaft. This is indicative of the presence of viscoelastic forces in the form of normal stresses. Such claims about the rheological properties of xanthan in simple solvent systems are usually not made in the commercial literature. It is possible that the calcium expected in high concentrations in Charles' whey based growth medium is interacting with xanthan, an interaction which is known[52, 53], to give a markedly viscoelastic solution. Tako et al.[53] studied the dynamic viscoelasticity of xanthan solutions and their results indicate that

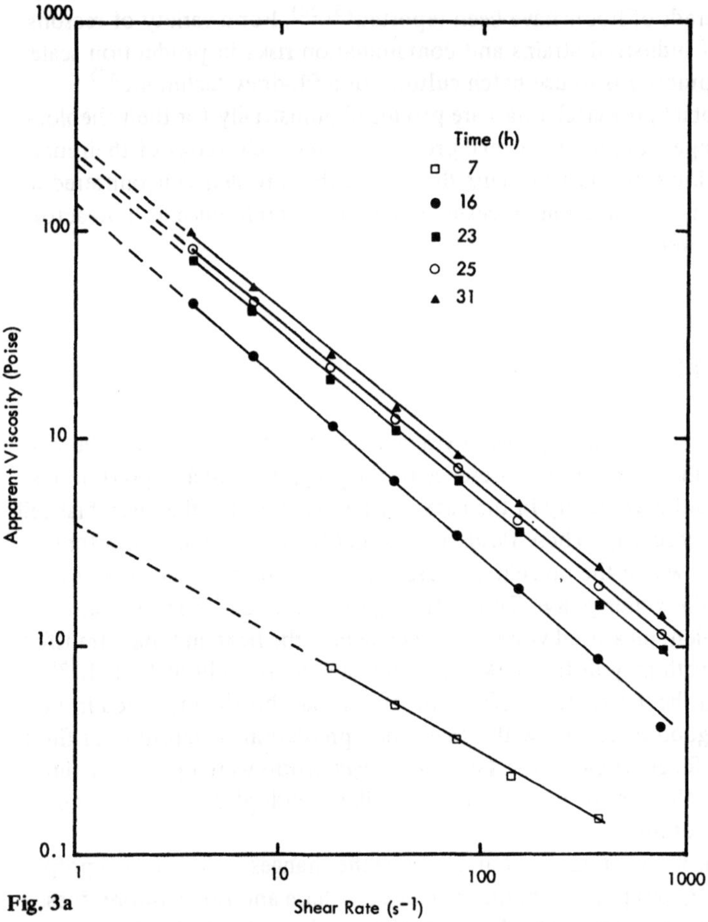

Fig. 3a

Fig. 3a and b. The change in culture viscosity over the course of a typical batch xanthan culture. The consistency index (K) and flow behaviour index (n) are defined by the equation: apparent viscosity = K (shear rate)$^{n-1}$

xanthans show elastic properties and are similar to those of other polymer solutions such as some carboxy methyl celluloses[50)]. These measurements represent the solid-like properties which exist when the fluid is at rest or subjected to small deformations and should not be confused with steady shear measurements of viscosity and normal stresses.

6 Reactor Design

6.1 Introduction

The maximum productivity and product concentration achievable in industrial microbial processes for the production of biomass or products is typically limited by the rate of

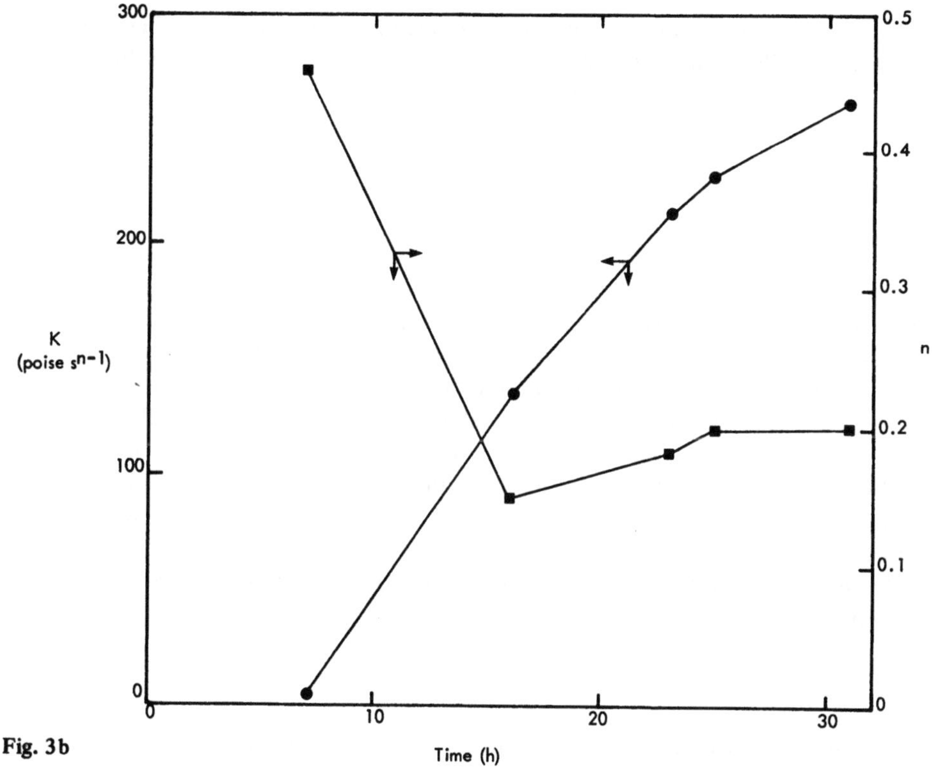

Fig. 3b

transfer of oxygen or heat removal, which is determined by the design of the mixing system and vessel used and by the fluid properties. The rheology of the culture fluid is an important factor in determining the heat and mass transfer capabilities of bioreactors and is most extreme in microbial processes developed for the production of extracellular microbial polysaccharides.

Conventional, mechanically agitated, aerated tanks appear to be widely used in polysaccharide production. Air agitated vessels appear unsuited to polysaccharide production due to the difficulty in promoting the necessary liquid turbulence for small bubble formation at the high liquid viscosities encountered with air sparging alone[54]. The use of a "thin film" reactor to achieve high productivities, through improved mass transfer is described by Lipps[55]. In his process *Xanthomonas campestris* is grown on a low carbohydrate medium in a conventional stirred tank. The resulting inoculum, which contains little polysaccharide, and a high carbohydrate medium are continuously applied as a thin film to a moving surface over which air is passed, and after a given period of time the polysaccharide containing broth is scraped from the surface. However, this approach does not appear to have been adopted in the industry. The proper design of an aerated, stirred reactor relies on achieving the correct balance between impeller velocity head, which controls gas bubble formation, and impeller flow, which controls the distribution of gas bubbles and liquid homogeneity throughout the tank and heat transfer rates.

The relationships between mixing and heat and mass transfer in microbial polysaccharide culture fluids and related non-Newtonian viscous fluids and the way they relate to reactor agitator design and scale-up is discussed below.

6.2 Power Consumption

The power input into a vessel and its method of dissipation is important in determining the liquid homogeneity and the heat and mass transfer characteristics of the reactor. The popular methods for estimating power consumption by agitators in unaerated non-Newtonian liquids in the laminar and transition regimes use expressions derived by analogy with viscometric or pipeline theory, relating the average impeller shear rate to one or more of the following: impeller and vessel dimensions, impeller speed, fluid properties[56, 57]. An average apparent viscosity can then be obtained using this relationship, and, by knowing the viscometric properties of the liquid being agitated, the power consumption for a particular impeller can be calculated from the power verses Reynolds number curve for that impeller.

Metzner and co-workers[58, 59] found a linear relationship between the average impeller shear rate ($\dot{\gamma}_{av}$) and impeller speed (N).

$$\dot{\gamma}_{av} = BN \, . \tag{1}$$

For example, they found that the constant B = 11 ± 1.4 for a six-flat-bladed turbine impeller was independent of fluid rheology for pseudoplastic and Bingham plastic fluids in either laminar and transitional flow and was valid for impeller to tank diameter ratios form 0.18–0.77. The values of B for other impellers and fluids are given in references[56] and[60] and Metzner and Otto's procedure for determining B is summarized in references[51, 61].

Calderbank and Moo-Young[62] using an analogy between shear in pipelines and in stirred vessels derived the following expression for the average impeller shear rate in power law fluids:

$$\dot{\gamma}_{av} = BN \left(\frac{4\,n}{3\,n+1}\right)^{n/1-n} . \tag{2}$$

For a six-flat-bladed turbine with an impeller to tank diameter ratio of less than 0.33 in a fluid with a consistency index (n) of 0.55–1.68 Eq. (2) reduces to:

$$\dot{\gamma}_{av} = 13.3\ N \pm 11.5\% \tag{2a}$$

and is similar to Metzner's expression.

Ranade and Ulbrecht[63] found that in the transition regime solutions of viscoelastic polymers showed a lower power consumption than solutions of inelastic polymers and hypothesised that this may be the cause of some of the scatter in results in power/

Reynolds number curves present in the literature[56]. The decrease in the power drawn by an impeller on aeration is a function of impeller and sparger geometry and size, air flow rate (Q) and fluid properties[60, 64]. The two correlations relating gassed to ungassed power (Pg/P_0) most widely quoted in the literature are:
Aeration Number Correlation[65]

$$Pg/P_0 = f(Q/ND^3)\,. \tag{3}$$

Michel Miller Correlation[66]

$$Pg = K\left(\frac{P_0^2\,ND^3}{Q^{0.56}}\right)^{0.45}\,. \tag{4}$$

Neither correlation takes into account fluid rheology and investigators have found that these do not adequately predict gassed power consumption in viscous systems[67–69].

The reduction in power consumption on aeration has recently been shown to be due to the formation of gas cavities behind the stirrer blades[64, 68]. The number, size and shape of the gas cavities was found to be a function of impeller configuration and dimensions, air flow rate, stirrer speed and liquid properties, notably viscosity.

The formation of large cavities which totally blanket the trailing side of the stirrer blades corresponds to the point at which the major reduction in power consumption is observed. In agitated viscous systems it has been observed that stable gas cavities remained behind the stirrer blades even after the air had been turned resulting in a Pg/P_0 of less than 1. These cavities which persisted in the absence of sparged air became less stable when either the viscosity or stirrer speed was decreased[63, 67]. Ranade and Ulbrecht[63] studying the power consumption in aerated viscoelastic fluids found that at a fixed gas flow rate and when a minimum stirrer was exceeded, large stable cavities abruptly formed behind all stirrer blades and accompanied a large decrease in power consumption. Beyond this point they found that varying the gas flow rate at a fixed stirrer speed had no effect on power consumption. This sharp discontinuity in the Pg/P_0 vs Q/ND^3 curve for viscous liquids compared to water has also been observed by others[69].

6.3 Mixing and Impeller Design

The turbulence and shear rate in a liquid in which a straight bladed turbine is rotating is much higher in the region of the impeller than in the bulk of the liquid[70]. In highly viscous pseudoplastic liquids this effect is enhanced due to the rapid increase in viscosity in the fluid at points moving away from the impeller and can lead to marked inhomogeneity. Presumably, increasing the yield stress of a solution will also adversely affect homogeneity.

The marked viscosity gradient which can exist in stirred pseudoplastic polysaccharide cultures is likely to promote channelling of the air up the centre of the vessel. Thus the optimal impeller design provides for sufficient turbulence to give good gas disper-

the diffusivity of low molecular weight solutes. The diffusivity of low molecular weight solutes such as dissolved gases in polymer solutions appears to be approximately constant over wide changes in the rheology of polymer solutions[84], and in some instances the diffusivities are increased with increased polymer concentration[85]. Thus, it is unlikely that a decrease in the diffusivity of oxygen or other nutrients is likely to be a problem in polysaccharide cultures. The behaviour of bubbles in the Stokes regime[86], spherical bubbles with rigid interfaces moving with creeping motion ($N_{Reb} < 1$), should correlate best to a well dispersed gas phase in a highly viscous polysaccharide liquid. Both experimental and theoretical work for a single bubble moving in Stokes regime, and for a single bubble with a free interface in creeping motion, show that the correction factors, F for the drag coefficient (C_D) [Eq. (7)] and F_{sh} for the Sherwood number [Eqs. (8 a) and (8 b)], increase with increasing pseudoplasticity or yield stress compared to a Newtonian Fluid[87, 91]

$$C_D = \frac{24\,F}{N_{Reb}} . \tag{7}$$

Immobile interface

$$N_{sh} = k\,F_{sh}\,Np_e^{0.33} . \tag{8 a}$$

Mobile interface

$$N_{sh} = k'\,F'_{sh}\,N_{Pe}^{0.5} . \tag{8 b}$$

Acharya et al.[90] have also shown that in this regime where viscous forces dominate, elasticity makes no significant contribution to the drag coefficient correction factor.

Bhavaraju et al.[92] in a theoretical study on the motion of swarms of bubbles in pseudoplastic fluids with either free or rigid interfaces predicted that the correction factors for the drag coefficient and Sherwood number decrease with increasing pseudoplasticity compared to a Newtonian fluid, the opposite to the single bubble case. They suggest that the retarding influence of neighbouring gas bubbles is less in the case of pseudoplastic liquids compared to Newtonian liquids of similar viscosity because of shear thinning of the liquid between the bubbles, i.e., lower apparent viscosities, and hence less resistance to motion or drag. Thus the increase in pseudoplasticity with time typically observed in microbial polysaccharide culture fluids may result in decreased mass transfer because of the lower drag and hence a decrease in the drag and mass transfer correction factors (F and F_{sh}).

In poorly mixed, highly viscous microbial cultures in which large bubbles are present the probable regime of interest is the Taylor regime, i.e., large spherical cap-shaped bubbles with a free interface and high bubble Reynolds numbers ($N_{Reb} \gg 1$)[86]. In this flow region the terminal rise velocity and mass transfer coefficient of a single bubble are insensitive to changes in rheology[86, 90, 93].

The size and distribution of bubbles is governed by the balance between liquid turbulence and interfacial forces and by the rate of coalescence of bubbles. In turbulent flow the maximum bubble size and distribution is controlled by the turbulent and interfacial forces with large bubbles formed by coalescence being rapidly broken up in the turbulent field. In laminar flow where the turbulent stresses are much less, both the maximum bubble size and distribution are strongly affected by coalescence. Thus increased viscosity results in lower Reynolds numbers, high rates of coalescence and a concomitant decrease in the gas/liquid interfacial area.

Bubble coalescence occurs in two independent steps, firstly the bubbles approach each other until separated by a thin film and secondly the inter bubble film drains, ruptures, and the bubbles coalesce. These two steps are characterized by a collision time and thin filming or coalescence time, respectively[94)].

At low Reynolds numbers ($N_{Reb} < 1$), the bubble enters the trailing wake of the leading bubble and due to the lower drag in the wake, accelerates towards the leading bubble. The collision time between the bubbles is decreased in viscous liquids because a longer wake exists behind bubbles moving in these liquids, and hence trailing bubbles tend to be caught in the wake of leading bubbles at a much greater distance than those in lower viscosity liquids[95)]. In contrast at low Reynolds numbers increasing the viscoelasticity of the liquid increases the collision time between two bubbles[94)]. At high Reynolds number viscous and viscoelastic stresses are comparatively small and have little or no effect on the collision time and the size of bubbles is governed by the magnitude of the turbulent fluctuations.

The draining and subsequent rupturing of the inter bubble film occurs when it is less than a few microns in thickness[96)]. The interbubble film can be assumed to be in laminar flow even though the liquid flow may be turbulent and thus the coalescence time is increased by increased liquid viscosity. It has also been found that increased viscoelasticity increases the coalescence time and thus decreases the rate of coalescence[94)]. In turbulent flow the time which the turbulent eddies hold the bubbles together compared to the coalescence time determines whether the bubbles coalesce or not[97)].

6.5 Heat Transfer

Heat transfer from unaerated, non-Newtonian liquids is comprehensively covered in reviews by Edwards and Wilkinson[98, 99)] and Astarita and Mashelkar[84)]. Correlations used to predict the process side heat transfer coefficient in unaerated, mechanically agitated, baffled, vessels containing non-Newtonian liquids are typically of the form[99)]

$$N_{Nu} = A N_{Re}^{q} N_{Pr}^{b} N_{Vi}^{c} \,. \tag{9}$$

An apparent viscosity, calculated from an impeller speed related average shear rate (see Power Consumption), is typically used in the Reynolds and Prandtl numbers, using this approach the viscosity number for a pseudoplastic liquid becomes:

$$N_{Vi} = \frac{K}{K_w} \,. \tag{10}$$

There is reasonable agreement between investigators on the values of the exponents in Eq. (9) with a = 0.67 (falling to less than 0.6 for laminar and transitional flow), b = 0.33, and c = 0.14[99]. The value of A varies widely and is a composite value being a function of impeller and vessel geometry. As published data appear restricted to studies on tanks with diameters of less than 1.2 m and usually less than 0.7 m care should be exercised in using these correlations in scale-up, particularly as the bulk liquid motion at a constant Reynolds number changes with scale, and the Reynolds number used in these correlations is related to the fluid in the region of the impeller rather than at the heat transfer surfaces.

Heat transfer in aerated agitated vessels containing Newtonian and non-Newtonian liquids has received little attention in the literature. Rao and Murti[100] found the process side heat transfer coefficient for aerated non-viscous and viscous Newtonian liquids was correlated by an expression similar to Eq. (9), in which a modified Reynolds number containing a superficial gas velocity (V_s) term was used. They found the heat transfer coefficient increased with V_s, the increase being greatest at low stirring speed. In contrast, Pollard and Topiwala[101] found that in a water system at low stirring speeds the heat transfer coefficient increased to a small extent with V_s and at higher stirring speeds was unaffected by changes in V_s. Pollard and Topiwala suggested their data was best correlated by an expression relating the heat transfer coefficient to the total power input (air compression plus agitator power) and V_s, Edney and Edwards[69] also found that the heat transfer coefficient for water, measured at various stirring speeds, was unaffected by V_s. The differences between these observations may be related to differences in the relative geometries of the cooling coils and impellers used in the vessels, and their effect on the bulk liquid motion at the various stirrer speeds and air flow rates studied.

Edney and Edwards also studied heat transfer into viscoelastic, pseudoplastic solutions of polyacrylamide and CMC and found under non-aerated conditions that their data was adequately correlated by an equation similar to Eq. (9), which only takes the pseudoplastic nature of the fluid into account. They found that as with heat transfer from water, on aeration of the polymer containing solutions the heat transfer coefficient, at a constant stirring speed, was not significantly affected by changes in V_s.

7 Recovery and Additional Processing of Extracellular Microbial Polysaccharides

Processes for recovering polysaccharides from microbial cultures are often based on isolation and purification methods developed for the production of gums from plants and seaweeds, examples of which are given in Table 4. Some of the extraction methods for releasing certain plants or algal gums from cellular materials have been applied to the extraction of cellular polysaccharides from micro-organisms, such as the hot alkali extraction of glycan from bakers yeast cell walls[102]. The recovery of extracellular microbial polysaccharides typically follows conventional plant and seaweed gum processing,

Table 4. Examples of unit operations used in the processing of plant and seaweed gums

Unit operation	Example (Reference)
A. *Extraction*	
1. Acid leaching	Arabinogalactan from wood[137], pectin from peel[138]
2. Alkali leaching	Alginate from brown algae[133], carrageenan from *Chondrus sp., Gigartina sp.*[135]
3. Irradiation of dried raw material	Irradiation of pectin containing peel or alginate containing seaweed to increase rate of acid or alkali extraction[139]
B. *Concentration purification separation*	
1. Concentration	Evaporation of carrageenan[135], ultrafiltration of carrageenan[140]
2. Solids removal	Centrifugation, filtration of carrageenan[135]
3. Colour removal	Resin treatment of agar[132], sulphite or hypochlorite bleaching of agar[132]
4. Salt precipitation	Pectin as aluminium salt[138], alginate as the calcium salt[133]
5. Drying of extract	Spray or drum drying of pectin[138] and carrageenan[135]
6. Electrolysis	Electrolytic desposition of alginic acid[134]
7. Freeze thawing	To remove salts and other soluble matter: furcellaran[136], agar[132]
8. Alcohol precipitation	Precipitation of carrageenan[135] or pectin[138] with methanol or isopropanol

and involves alcohol or salt precipitation of the polymer, solids separation, conversion of the precipitate to water soluble form, dewatering, drying, and milling. However, some of the more unusual methods listed in Table 4 which have been proposed or used in plant or seaweed processing may also find a niche in microbial gum recovery operations. Additional unit operations such as cell removal, destruction of undesirable enzyme activities and chemical or enzymatic modification of polysaccharides may also be employed. The processing methods of choice are determined by the application of the gum and to illustrate the variety of process options available, a review of methods proposed for the processing of xanthan with some references to other microbial polysaccharides is given below.

7.1 Recovery

The precipitation of polysaccharides is usually accomplished by the addition of a solvent, such as methanol, ethanol, isopropanol or acetone: the addition of specific ions which at certain pH's give insoluble salts of the polysaccharides; or by adjusting the pH to give the insoluble acid form. The firmness and fibrous nature of the precipitate is affected by the precipitation conditions and mechanical stresses during precipitation. This characteristic determines the ease of handling in later processing. Also, the colour of the product can be affected by the efficiency of solvent contact during precipitation.

Methods of precipitating xanthan include: solvent precipitation, e.g., addition of acetone, methanol, ethanol, isopropanol or 1,1,1-trichloroethane[103, 104)]; precipitation as the divalent metal ion salt of the polymer at alkaline pH's, e.g., as the calcium salt[105-109)]; precipitation as the aluminium salt at acid pH's[110)]; precipitation as a quaternary ammonium complex[111, 112)]; or precipitation as a fatty amine complex at acid pH's[113)]. Alternatively, the whole liquid can be spray or drum dried although in this case the final product is highly coloured and contains unmetabolised components[114)]. The method most favoured for the recovery of xanthan is solvent precipitation using approximately 45–60% by weight of isopropanol to precipitate the polymer[115)]. The operational costs of an alcohol based polysaccharide recovery process are largely determined by the amount of alcohol lost per weight of the product during processing and the alcohol distillation costs. The minimum amount of alcohol required to precipitate xanthan is dependent on ionic strenght and to a much lesser extent on polymer concentration in the broth (Table 5). As the amount of alcohol required to precipitate

Table 5. The effect of ionic strength and xanthan concentration on the amount of isopropanol required for xanthan precipitation

Concentration of potassium chloride (% w/w)	Concentration of xanthan (% w/w)	Volume of alcohol required per volume of xanthan solution
0	1.0	1.0
1	1.0	0.7
0	2.0	0.9
1	2.0	0.6

xanthan from a given volume of broth is not very dependent on the concentration of product in the broth, substantial savings in the amount of alcohol lost during processing and the distillation costs can be made by increasing the concentration of polymer in the broth, for example, by growing to higher product concentrations. The wet precipitate, which contains the bulk of the microbial cells, is then dewatered, dried, and milled. The final milled product from this process contains no viable *Xanthomonas campestris* and complies with the FDA description of a process for making food grade material.

A scheme proposed for the recovery of microbial alginates, based on algal alginate processing, is shown in Fig. 5. In this process, the microbial cells are first removed from the broth and the alginate is then precipitated by the addition of calcium chloride. The insoluble calcium alginate is separated, dewatered and then washed with acid to convert the calcium salt to the insoluble alginic acid. To produce a marketable product the alginic acid is titrated with solid sodium hydroxide and sodium carbonate to approximately pH 7–8 to give the water soluble sodium alginate. The wet solid is the dewatered, dried and milled.

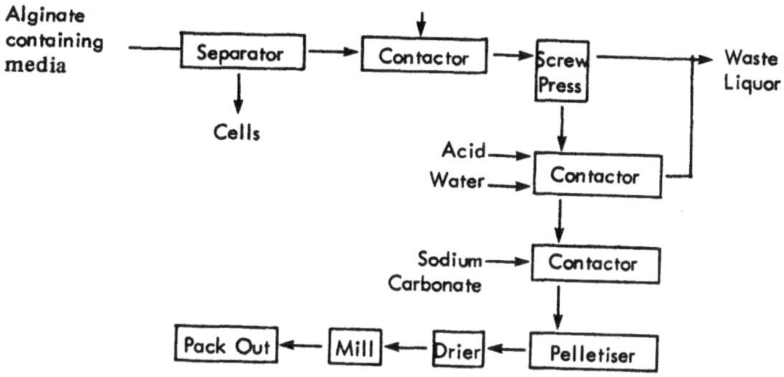

Fig. 5. Proposed process for purification and recovery of microbial alginate

Polysaccharide precipitates are usually dried in batch or continuous driers under vacuum, or using forced air or inert gas if the solids contain solvent. The proper choice of drying conditions is critical if chemical degradation, excessive colouring or changes in the solubility of the product are to be avoided. For example, microbial alginate is degraded when heated above 90° for more than about one hour with solutions of the final product showing lower visocsities. Rapid drying of polysaccharides can result in case hardening or hornification with the resultant milled product showing poor hydrability, i.e., a measure of the time required for a completely dispersed product to reach its final viscosity. Drying conditions can also affect the dispersibility, i.e., the tendency of the gum to form clumps when added to water, and bulk density of the product.

The dried polysaccharide is milled to a predetermined mesh size in a further attempt to control dispersibility and hydration rate and to make solids handling easier. Care must be taken during milling to avoid excessive heating which can lead to darkening or degradation of the polymer. Typically, polysaccharides are hydroscopic in nature and moisture picked up during transportation and storage can lead to clumping of the product in the container and sometimes accelerate the rate of product degradation during storage. Thus special attention is given to the water permeability characteristics of packaging materials as well as their mechanical strength.

7.2 Additional Processing

Additional operations to those already mentioned in the previous section may involve specific treatments to improve or modify the properties of the final product or remove unwanted impurities carried over from the culture stage. As with product isolation, modification, and purification procedure applied to microbial polysaccharides are often based on methods established for plant and seaweed gums.

Many industrial and food grade natural and synthetic gums are processed to remove water insoluble solids, thus making the product cosmetically attractive and handling of solutions easier. Numerous techniques for the removal of insoluble medium solids and cells from xanthan culture fluids have been proposed, and include: dilution of the medium followed by filtration or centrifugation, and sometimes in combination with solids or cell flocculation[116–120]; alkaline protease digestion of cells[121]; alkali degradation of cells[122] which may also result in deacetylation and degradation of the polymer; hydrochlorite degradation of cells[123], though severe oxidation may result in depolymerisation. Filtration or centrifugation of cells from xanthan can be improved by heating the liquid, but owing to the unusual viscosity temperature profile of xanthan gum solutions (Fig. 6), the solutions must be heated above the "melting point" of the xanthan[52], which depending on the ionic strength of the solution could be above 100 °C.

Polysaccharides typically degrade when heated at high temperatures, however, when media containing xanthan are heated under defined conditions, e.g., 80–130 °C for 10–120 min, the viscosity of the finished product increases[124]. The exact mechanism for this is unknown, although it is thought to involve aggregation of the polysaccharide. When heated at higher temperatures (e.g., above approx. 150 °C for 2 min) xanthan will also degrade.

The presence of certain enzymes in microbial polysaccharide cultures may cause problems in subsequent processing, or impart undesirable characteristics to the final product. The presence of alginate lyase can lead to severe degradation of the polymer[125] during production and recovery operations, and similarly loss of culture viscosity has been observed[47, 48] during the latter stages of batch pullulan cultures. To minimise such degradation, depolymerase free mutants or methods for inhibiting or destroying

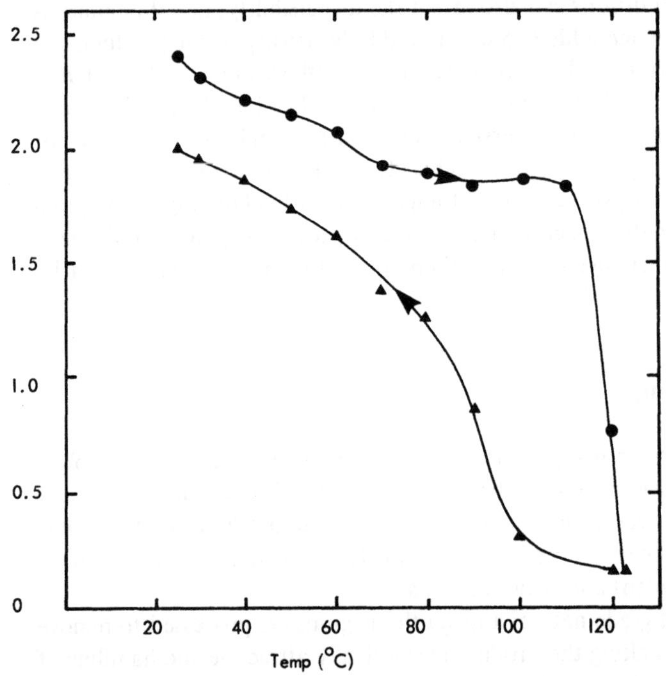

Fig. 6. The effect of heating and cooling on the apparent viscosity of a 1% xanthan in 0.5% sodium chloride

these enzymes are required. The presence of cellulase activity in some commercially available xanthan gums can lead to problems when blending xanthan with cellulose ethers, such as carboxymethyl cellulose; however, there is no evidence to indicate that the *Xanthomonas* cellulase is capable of degrading xanthan which has a cellulose backbone. One method proposed for destroying the cellulase activity present in xanthan gum is to treat the wet precipitate with propylene oxide, propiolactone, glutaraldehyde or pivalolactone and then air dry[126)].

The treatment of microbial polysaccharides with enzymes may find some utility in the modification of gums. The composition of microbial alginate can be altered by controlling the activity of an epimerase enzyme present in the culture. For example, the level of alginate epimerase in *A. vinelandii* culture fluids which catalyses the conversion of mannuronic acid to guluronic acid residues present in alginate, is increased by the addition of calcium ions[12, 18)]. By altering the ratio of mannuronic residues in the polymer the rheological properties of the finished products are changed. Also, a range of grades of the plant gum tamarind with different viscosities can be prepared by treating the isolated gum with a fungal cellulase[127)], and such approaches should be applicable to microbial polymers.

The physical and chemical properties of polysaccharides can be significantly altered by reacting them with a variety of chemicals, e.g., reaction of xanthan with glyoxal yields a product which is easily dispersed in water, i.e., it does not clump[128)]. The glyoxal xanthan complex is initially insoluble when added to water the glyoxal-xanthan crosslinks slowly hydrolyse allowing the xanthan to dissolve. When xanthan is heated at alkaline pH's, e.g., pH 11 for 30 min, the polymer is deacetylated and the viscosities of solutions of the resultant product are claimed to be enhanced by the addition of salt[129)]. Crosslinking xanthan with formaldehyde gives a stable, water soluble polymer with enhanced viscosity[130)]. Also, Stratton[131)] described a method for forming a graft polymer between polyacrylamide and xanthan, which is claimed to have superior properties compared to xanthan when used as an additive in polymer floods in enhanced oil recovery.

8 Concluding Remarks

The design of economically viable processes for the production of extracellular microbial polysaccharides presents several interesting challenges to the microbiologist, engineer, and microbial technologist. Some fruitfull areas for research include: devising methods for altering and controlling the composition and molecular weight of polymers, for example, by manipulating culture conditions or mutation and selection techniques; design reactors with improved heat and mass transfer capabilities at high product concentration by possibly using the extreme pseudoplasticity of shear thinning of the broth to advantage; determine the relationships between fundamental studies on heat and mass transfer in model systems and the behaviour observed in aerated microbial polysaccharide culture fluids; and economic methods of product isolation and cell removal.

9 Nomenclature

A, b, c, k, k′, k	constants
k	thermal conductivity
k_L	mass transfer coefficient
k_La	overall volumetric mass transfer coefficient
g	gravitational constant
h	heat transfer coefficient
n	flow behaviour index
q	impeller pumping capacity
A, B	constants
C_D	drag coefficient = $8\,gR/3\,U^2$
C_p	specific heat
D	impeller diameter
D_T	tank diameter
D_L	molecular diffusivity
F	correction factor for drag coefficient [Eq. (7)]
F_{Sh}	correction factor for mass transfer [Eq. (8a)]
F'_{Sh}	correction factor for mass transfer [Eq. (8b)]
H	impeller velocity head
K	consisteny index
K_w	consistency index at wall temperature
N	impeller speed
N_{Nu}	Nusselt number = $hD_{T/k}$
N_{Pe}	Peclet number = $2\,RU/D_L$
N_{Pr}	Prandtl number = $\mu\,C_{p/k}$
N_{Re}	Reynolds number – impeller = $\rho\,ND^2/\mu$
N_{Reb}	Reynolds number-bubble = $(2\,R)^n = U^{2-n}\,\rho/K$
N_{Sh}	Sherwood number = $2\,k_LR/D_L$
N_{Vi}	viscosity ratio = μ/μ_ω
P	power
P_0	ungassed power
Pg	gassed power
R	bubble radius
U	bubble velocity
V_s	superficial gas velocity
$\dot{\gamma}_{av}$	average impeller shear rate
ρ	density
μ	apparent viscosity
μ_ω	apparent viscosity at wall temperature

10 References

1. Sutherland, I.W.: Surface carbohydrates of the prokaryotic cell, p. 27. London: Academic Press 1977
2. Jarman, T.R., Deavin, L., Slocombe, S., Righelato, R.C.: J. Gen. Microbiol. *107*, 59 (1978)

3. Laskin, A., Lechevalier, H.: Handbook microbiology, condensed Edit., p. 680. Cleveland, Ohio: C. R. C. 1974
4. Johnson, J., Kirkwood, S., Misaki, A., Nelson, T.E., Scaletti, J.V., Smith, F.: Chem. Ind., p. 820, 1963
5. Harada, T.: Fermentation technology today: Proc. 4 th Internat. Fermentation Symposium, p. 603, 1972
6. Jansson, P.E., Keene, L., Lindberg, B., Ljunggren, H., Lonngren, J., Ruden, U., Svensson, S.: J. Am. Chem. Soc. *11*, 3812 (1977)
7. Cadmus, M.C., Knutson, C.A., Lagoda, A.A., Pittsley, J.E., Burton, K.A.: Biotech. Bioeng. *20*, 1003 (1978)
8. Davidson, I.W., Sutherland, I.W., Lawson, C.J.: J. Gen. Microbiol. *98*, 603 (1977)
9. Rees, D.A.: Chem. and Ind., p. 630 (1972)
10. Deavin, L., Jarman, T.R., Lawson, C.J., Righelato, R.C.: In: Extracellular microbial polysaccharides. Sandford, P.A., Laskin, A. (eds.), p. 14. Am. Chem. Soc., Washington, D.C. 1977
11. Gibbons, R.J., Nygaard, M.: Arch. Oral Biology *13*, 1249 (1968)
12. Lawson, C.J., Sutherland, I.W.: Economic microbiology, Vol. 2. New York: Academic Press 1978
13. Bucke, C.: Chromatography *89*, 99 (1974)
14. Sutherland, I.W.: Soc. Gen. Microbiol. Proc., p. 107, 1978
15. Holzwarth, G.: Carbohydrate Res. *66*, 171 (1978)
16. Jeanes, A.R., Sloneker, J.H.: U.S. Pat. 3,054, 689 (1962)
17. Sandford, P.A., Pittsley, J.E., Knutson, C.A., Watson, P.R., Cadmus, M.C., Jeanes, A.: In: Extracellular microbial polysaccharides. Sandford, P.A., Laskin, A. (eds.), p. 192. Am. Chem. Soc., Washington D.C. 1977
18. Haug, A., Larsen, B.: Carbohydrate Res. *17*, 297 (1971)
19. Larsen, B., Haug, A.: Carbohydrate Res. *17*, 287 (1971)
20. Pindar, D.F., Bucke, C.: Biochem. J. *152*, 617 (1975)
21. Sutherland, I.W.: In: Extracellular microbial polysaccharides. Sandford, P.A., Laskin, A. (eds.), p. 40. Am. Chem. Soc., Washington, D.C. 1977
22. Neijssel, O.M., Tempest, D.W.: Arch. Microbiology *107*, 215 (1976)
23. Williams, A.G.: Extracellular polysaccharide production by a gram-negative bacterial isolate. Ph. D. thesis, University of Wales 1974
24. Silman, R.W., Rogovin, S.P.: Biotech. Bioeng. *14*, 23 (1972)
25. Mian, F., Jarman, T.R., Righelato, R.C.: J. Bact. *134*, 418 (1978)
26. Davidson, I.W.: FEMS Microbiology Lett. *3*, 347 (1978)
27. Williams, A.G., Wimpenny, J.W.T.: J. Gen. Microbiol. *104*, 47 (1978)
28. Ellwood, D.C., Evans, C.G.T., Yeo, R.G.: U.K. Pat. 1512 536 (1978)
29. Harada, T., Yoshimura, T., Hidaka, H., Koreeda, A.: Agricult. Biol. Chem. *29*, 757 (1965)
30. Evans, L.R., Linker, A.: J. Bact. *116*, 915 (1973)
31. Wilkinson, J.F., Duguid, J.P., Edmunds, P.N.: J. Gen. Microbiol. *11*, 59 (1954)
32. Silman, R.W., Rogovin, P.: Biotech. Bioeng. *12*, 75 (1970)
33. Moraine, R.A., Rogivin, P.: Biotech. Bioeng. *15*, 225 (1973)
34. Godet, P.: Proc. Biochemistry *8* (1), 33 (1973)
35. McNeely, W.J.: U.S. Pat. 3391060 (1969)
36. Lindblom, G.P., Patton, J.: U.S. Pat. 3328262 (1967)
37. Tong, G.: Int. Biochem. Symp. Stone and Webster Canada Limited 1977
38. Jeanes, A.: Food Technol. *28*, 34 (1974)
39. Jeanes, A.: J. Polymer Sci.: Symposium No. 45, 209 (1974)
40. Catley, B.J.: FEBS Letters *20*, 174 (1972)
41. Hofheinz, W., Grisebach, H.: Chem. Ber. *96*, 2867 (1963)
42. Adams, G.A., Martin, S.M.: Can. J. Biochem. *42*, 1403 (1964)
43. Simley, K.L.: Food Technol. *20*, 112 (1966)
44. Spencer, J.F.T., Gorin, P.A.J.: Biotechnol. Bioeng. *15*, 1 (1973)

45. Farr, D.R., Armister, H., Horisberger, M.: Carbohydrate Res. *24*, 207 (1972)
46. Blakebrough, N., McMananey, W.J., Tart, K.R.: Trans. IChemE, *56*, 127 (1978)
47. Leduy, A., Marsan, A.A., Coupal, B.: Biotechnol. Bioeng. *16*, 61 (1974)
48. Miura, Y., Fukushima, S., Sambuichi, M., Ueda, S.: J. Ferment Technol. *54*, 166 (1976)
49. Whitcomb, P.J., Ek, B.J., Macosko, C.W.: In: Extracellular microbial polysaccharides. Sandford, P.A., Laskin, A. (eds.), p. 160. Am. Chem. Soc., Washington D.C. 1977
50. Elliott, J.E.: In: Extracellular microbial polysaccharides. Sandford, P.A., Laskin, A. (eds.), p. 144. Am. Chem. Soc., Washington D.C. 1977
51. Charles, M.E.: In: Adv. Biochem. Eng. Ghose, T.K., Fiechter, A., Blakebrough, N. (eds.), Vol. 8, p. 1. Berlin, Heidelberg, New York: Springer 1978
52. Holzwarth, G.: Biochemistry *15*, 4333 (1976)
53. Tako, M., Nagahama, T., Nomura, D.: J. Agric. Chem. Soc. (Japan) *51*, 513 (1977)
54. Bhavaraju, S.M., Russell, T.W.F., Blanch, H.W.: AIChEJ *24*, 454 (1978)
55. Lipps, B.J.: U.S. Pat. 3,281,329 (1966)
56. Bates, R.L., Fondy, P.L., Fenic, J.G.: Mixing. Uhl, V.W., Gray, J.B. (eds.), Vol. 1. New York: Academic Press 1966
57. Holland, F.A., Chapman, F.S.: Liquid mixing and processing in stirred tank. New York: Reinhold 1966
58. Metzner, A.B., Feehs, R.H., Ramos, H.L., Otto, R.E., Tuthill, J.D.: AIChEJ *7*, 3 (1961)
59. Metzner, A.B., Otto, R.E.: AIChEJ *3*, 3 (1957)
60. Nagata, S.: Mixing principles and applications. New York: John Wiley and Sons 1975
61. Aiba, S., Humphrey, A.E., Millis, N.F.: Biochem. Eng. New York: Academic Press 1973
62. Calderbank, P.H., Moo-Young, M.B.: Trans. IChemE *99*, 337 (1961)
63. Ranade, V.R., Ulbrecht, J.: Proc. Soc. Europ. Conf. on Mixing. Paper F6 Cambridge, England 1977
64. Van't Riet, K.: Turbine agitation hydrodynamics and dispersion performance. PhD Thesis, Delft Univers. 1975
65. Oyama, Y., Endoh, K.: Chem. Eng. (Japan) *19*, 2 (1955)
66. Michel, B.J., Miller, S.A.: AIChEJ *8*, 262 (1962)
67. Taguchi, H., Miyamoto, S.: Biotech. Bioeng. *8*, 43 (1966)
68. Bruijn, W., Van't Riet, K., Smith, J.M.: Trans. IChemE *52*, 88 (1974)
69. Edney, H.G.S., Edwards, M.F.: Trans. IChemE *54*, 160 (1976)
70. Oldshue, J.: Biotech. Bioeng. *8*, 3 (1966)
71. Uhl, V.W., Gray, J.B.: Mixing. Vols. 1 and 2. New York: Academic Press 1966
72. Johnson, D.N., Hubbard, D.W.: Biotech. Bioeng. *16*, 1283 (1974)
73. Hubbard, D.W., Williams, C.N.: Process Biochem. *12* (10), 11 (1977)
74. Giacobbe, F., Capobianco, G.: Paper presented Soc. Gen. Microbiol. Meeting. Reading, England, Sept. 1976
75. Moraine, R.A., Rogovin, P.: Biotech. Bioeng. *15*, 225 (1973)
76. Giesekus, H.: Rheol Acta *4*, 85 (1965)
77. Ulbrecht, J.: The chem. Eng., p. 347. London 1974
78. White, J.L., Chankraiphon, S., Ide, Y.: Trans. Soc. Rheol. *21*, 1 (1977)
79. Chavan, V.V., Arumugan, M., Ulbrecht, J.: AIChEJ *21*, 613 (1975)
80. Paca, J., Ettler, P., Gregr, V.: J. Appl. Chem. Biotech. *26*, 309 (1976)
81. Hattori, K., Yokoo, S., Imada, O.: J. Ferment: Technol. *50*, 737 (1976)
82. Perez, J.F., Sandall, O.C.: AIChEJ *20*, 770 (1974)
83. Yagi, H., Yoshida, F.: Ind. Eng. Chem. Process Des. Dev. *14*, 488 (1975)
84. Astarita, G., Mashelkar, R.A.: The Chem. Eng., p. 100, 1977
85. Mashelkar, R.A., Soylu, M.A.: Chem. Eng. Sci. *29*, 1089 (1974)
86. Astarita, G., Apuzzo, G.: AIChEJ *11*, 815 (1965)
87. Nakano, Y., Tien, G.: AIChEJ *14*, 145 (1968)
88. Hirose, T., Moo-Young, M.: Can. J. Chem. Eng. *47*, 265 (1969)
89. Wellek, R.M., Huang, C.C.: Ind. Eng. Chem. Fundam. *9*, 480 (1970)

90. Acharya, A., Mashelkar, R.A., Ulbrecht, J.: Chem. Eng. Sci. *32*, 863 (1972)
91. Bhavaraju, S.M., Mashelkar, R.A., Blanch, H.W.: AIChEJ *24*, 1063 (1978)
92. Bhavaraju, S.M., Mashelkar, R.A., Blanch, H.W.: AIChEJ *24*, 1070 (1978)
93. Calderbank, P.H., Johnson, D.S.L., London, J.: Chem. Eng. Sci. *25*, 235 (1970)
94. Acharya, A., Ulbrecht, J.J.: AIChEJ *24*, 348 (1978)
95. deNevers, N., Wu, J.L.: AIChEJ *17*, 182 (1971)
96. Narayanan, S., Goossens, L.H.J., Kossen, N.W.F.: Chem. Eng. Sci. *29*, 2071 (1974)
97. King, M.L., Russell, T.W.F., Blanch, H.W.: Paper presented at Meeting of Am. Chem. Soc., Miami, Florida, Sept. 11, 1978
98. Edwards, M.F., Wilkinson, W.L.: The Chem. Eng., p. 310. London 1972
99. Edwards, M.F., Wilkinson, W.L.: The Chem. Eng., p. 328. London 1972
100. Rao, K.B., Murti, P.S.: Ind. Eng. Chem. Process Des. Develop. *13*, 190 (1973)
101. Pollard, R., Topiwala, H.H.: Biotech. Bioeng. *18*, 1517 (1976)
102. Sucher, R.W., Robbins, E.A., Schuldz, E.H., Seeley, R.D., Newell, J.A.: 33rd Ann. Meeting of Institute of Food Technologists Miami, Florida 1973
103. Jeanes, A.: In: Encyclopedia of polymer science and technology, Vol. 8. New York: John Wiley and Sons 1968
104. Leader, H.J., Miescher, G.M.: U.S. Pat. 3,316,241 (1967)
105. Mehltretter, C.L.: Biotech. Bioeng. *7*, 171 (1965)
106. McNeely, W.J.: U.S. Pat. 3,232,929 (1966)
107. Patton, J.H., Holman, W.F.: U.S. Pat. 3,382,229 (1968)
108. Falcoz, P., Celle, P., Campagne, J.C.: French Pat. 2,299,366 (1976)
109. Cahalan, P.T., Peterson, J.A., Arudt, D.A.: U.S. Pat. 4,053,699 (1977)
110. Towle, G.A.: U.S. Pat. 4,051,317 (1977)
111. Albrect, W.J., Sohns, V.E., Rogovin, S.P.: Biotech. Bioeng. *5*, 91 (1963)
112. Rogovin, S.P., Albrecht, W.J.: U.S. Pat. 3,119,812 (1964)
113. Jordan, W.A., Carter, W.H.: U.S. Pat. 3,928,316 (1975)
114. Rogovin, P., Albrecht, W., Sohns, V.: Biotech. Bioeng. *7*, 161 (1965)
115. Bouniot, A.: U.S. Pat. 3,988,313 (1976)
116. Abdo, M.K.: U.S. Pat. 3,711,462 (1973)
117. Buchanan, B.B., Cottle, J.E.: Pat. 3,773,752 (1973)
118. O'Connell, J.: U.S. Pat. 3,355,447 (1967)
119. Patton, J.T., Holman, W.E.: U.S. Pat. 3,208,526 (1965)
120. Rogovin, S.P., Anderson, R.F., Cadmus, M.C.: Biotech. Bioeng. *3*, 51 (1961)
121. Colegrove, G.T.: U.S. Pat. 3,966,618 (1976)
122. Patton, J.T.: U.S. Pat. 3,964,972 (1976)
123. Colegrove, G.T.: U.S. Pat. 3,516,983 (1970)
124. Colin, P., Guibert, V.: U.S. Pat. 3,591,578 (1971)
125. Larsen, B., Haug, A.: Carbohydrate Res. *17*, 287 (1971)
126. Empey, R.A.: U.S. Pat. 4,070,535 (1978)
127. Rao, P.S., Srivastava, S.: In: Industrial gums. Whistler, R.L. (ed.). New York: Academic Press 1973
128. Maske, F.J.: U.S. Pat. 4,041,234 (1977)
129. Jeanes, A.R., Sloneker, J.H.: U.S. Pat. 3,096,293 (1963)
130. Patton, J.T.: U.S. Pat. 3,020,207 (1962)
131. Stratton, C.A.: U.S. Pat. 3,844,348 (1974)
132. Selby, H.W., Wynne, W.H.: In: Industrial gums. Whistler, R.L. (ed.). New York: Academic Press 1973
133. McNeely, W.H., Pettitt, D.J.: In: Industrial gums. Whistler, R.L. (ed.). New York: Academic Press 1973
134. Taiho Industrial Co., Ltd.: Brit. Pat. 856,416, 1962; Chem. Abstr. *62*, 4654, 1965
135. Towle, G.A.: In: Industrial gums. Whistler, R.L. (ed.). New York: Academic Press 1973

136. Bjerre-Petersen, E., Christensen, J., Hemmingsen, P.: Industrial gums. Whistler, R.L. (ed.). New York: Academic Press 1973
137. Adams, M.F., Ettling, B.V.: In: Industrial gums. Whistler, R.L. (ed.). New York: Academic Press 1973
138. Towle, G.A., Christensen, O.: In: Industrial gums. Whistler, R.L. (ed.). New York: Academic Press 1973
139. Lowry, J.R.: U.S. Pat. 3,034,973 (1962)
140. Strong, C.H.G.: U.S. Pat. 3,856,596 (1974)

Use of Immobilized Lactase in Milk Systems

Terry Finocchiaro, Norman F. Olson, Thomas Richardson
Department of Food Science, University of Wisconsin
Madison, WI 53706, USA

Immobilization of lactase for continuous hydrolysis of lactose in milk systems offers considerable potential for the improvement of fluid dairy products. The hydrolyzed product, which contains glucose and galactose, may possess improved functional and nutritional properties. In particular, the usage of cheese whey, a by-product of cheese manufacturing, may be expanded greatly with the development of immobilized lactase technology. Although lactases occur rather widely in nature, only microbial enzymes are of commercial value. The various approaches to lactase insolubilization have been reviewed. Additionally, pertinent factors in operating an immobilized lactase reactor system have been discussed. Commercial lactose reactors are being used currently for industrial production of low-lactose skimmed milk and appear to have economic potential for processing of cheese whey.

1 Introduction

Lactose [4-O-(β-D-galactopyranosyl)-D-glucopyranose] is the principal carbohydrate of milk, occuring at a concentration of about 5% w/v. It is hydrolyzed to equimolar concentrations of glucose and galactose by enzymes termed β-galactosidases (EC. 3.2.1.23), commonly referred to as lactases. Since cow's milk and various products made from it are major food items, lactose constitutes a significant portion of the daily carbohydrate intake of humans.

Lactose has many nutritional and functional attributes but its presence in milk and cheese whey can present problems to the processor and certain consumers of milk and other dairy products. Intolerance to lactose because of a deficiency of lactase in the gastrointestinal tract occurs in a segment of the population. The resultant excess lactose in the gut is metabolized by the microbial flora in the colon leading to diarrhea, cramps, flatulence and general intestinal discomfort. This problem can be circumvented if lactose in the product is hydrolyzed by lactase to the readily utilizable sugars, glucose and galactose. Insolubility of lactose can result in its crystallization in dairy foods causing defects such as sandy or gritty texture, deposit formation and protein destabilization. Products which suffer most from such crystallization are condensed milk, sweetened condensed milk, condensed and dried wheys, ice cream, and frozen milks. Using lactase in such products could reduce lactose concentration to where it no longer presents a problem. Presence of lactose in cheese whey, a by-product of cheese manufacturing creates problems in utilization or disposal of this product. Cheese production in the United States in 1974 was approximately 1×10^9 kg and resulted in the formation of 1×10^{10} kg of whey, of which only one-half was further processed. Since lactose comprises about 70% of the 6.5% (w/v) total solids of whey, it is an important factor in expanding the usage of whey. Enzymic hydrolysis of the lactose in whey might promote such developments. Additional reasons for hydrolyzing lactose in dairy products include increased sweetness of glucose and galactose over lactose, increased osmotic pressure of the hydrolysis products, which is sometimes desirable, and increased ease of utilization of the hydrolysis products by various microorganisms used in food systems[1)].

Although strong mineral acids[2)], and ion exchange resins in the cation form[3)] can also be used to hydrolyze lactose, only enzymic hydrolysis by lactase will be discussed further. Lactases can be used in the soluble form but development of immobilized enzyme technology presents the possibility of binding enzymes to insoluble matrices for use in continuous or batch-type reactors. This review will be restricted to the utilization of immobilized lactase in milk systems and will supplement several reviews which have appeared recently on soluble and immobilized lactases and their uses in milk processing[4–6)].

2 Sources of Lactase

Beta-galactosidases occur rather widely in nature with potential plant, animal and microbial sources being listed in Table 1. Enzymes of plant and animal origin are of little commercial value but several microbial lactases are of technological interest. The major enzymes of commercial interest are those from *Kluyveromyces lactis, Kluyveromyces fragilis* and *Aspergillus niger.*

2.1 Properties of Microbial Lactases

Characteristics of the principal microbial lactases are given in Table 2. Fungal enzymes generally have acidic pH optima in the range of 2.5–5.4 whereas yeast lactases are most

Table 1. Possible sources of β-galactosidase[a]

Plants	
Peach	
Apricot	
Almond	
Kefir grains	
Tips of wild roses	
Alfalfa seed	
Coffee	
Animal organs	
Intestine	
Brain and skin tissue	
Yeast	
Kluyveromyces lactis (formerly *Saccharomyces lactis*)	
Kluyveromyces fragilis (formerly *Saccharomyces fragilis*)	
Candida pseudotropicalis	
Bacteria	
Eschericia coli	
Bacillus megaterium	
Thermus equaticus	
Streptococcus lactis	
Streptococcus thermophilus	
Lactobacillus bulgaricus	
Lactobacillus helveticus	
Fungi	
Neurospora crassa	
Aspergillus foetidus	
Aspergillus niger	
Aspergillus flavus	
Aspergillus oryzae	
Aspergillus phoenicis	
Mucor pucillus	
Mucor miehei	
Curvularia inaequalis	
Commercial preparations	
Aspergillus niger	Baxter Laboratories, Chicago, IL
	Dairyland Food Labs, Waukesha, WI
	Kyowa Hakko Kogyo Co., Japan
	Societe Rapidase, Seclin, France
	Wallerstein Co., Morton Grove, IL
Kluyveromyces lactis	Gist-Brocade, Delft, Holland
	Nutritional Biochemical Co., Cleveland, OH
	Tokyo Tanabe Co., Ltd., Tokyo, Japan
Kluyveromyces fragilis	Kyowa Hakko Kogyo Co., Japan
	Sigma Chemical Co., St. Louis, MO
Escherichia coli	C.F. Boehringer GmbH, Mannheim, Germany
	Worthington Biochemical Corp., Freehold, NJ
Yeast Preparation	British Drug House Ltd., London
	DEBI, Cassina De Pecchi, Milan, Italy

[a] Adapted from original table by Shulka, T.P.
β-galactosidase technology: a solution to the lactose problem
CRC-Food Technology. *5*, 325 (1975)

Table 2. Properties of *A. niger* and *K. lactis* Lactases[a]

	A. niger	*K. lactis*
pH-Optimum	4.0–4.5	6.8–7.0
Temperature optimum	55 °C	35 °C
pH-Stability	3.0–7.0	6.0–8.5

[a] Adapted from original table by Woychik, J.H. and Holsinger, V.H. In: Ory, R.L., Angelo, A.J.St. (eds.). Enzymes in food and microbial processing. ACS Symposium Series, Washington, D.C. 1977

active between pH values of 6.0–7.0. Consequently, fungal enzymes should be most effective in acidic products such as whey, from manufacture of cottage cheese, with a pH value of 4.4–4.6. The yeast enzymes would be most useful in treating products (such as fluid milk) near a neutral pH.

Although lactases can be used conveniently in soluble mode, the economics of processing and potential for continuous processing require evaluation of immobilized beta-galactosidases as a means for continuously hydrolyzing lactose in fluid dairy products. However, it is possible that the soluble enzyme would be the only practical means of processing. For example Woychik et al.[1] have indicated that immobilized lactase from *K. lactis* is insufficiently stable to warrant its use for processing dairy products. Furthermore the pH of milk (where the enzyme is most active) is appropriate for the growth of a wide variety of microorganisms making sanitary use of the immobilized enzyme difficult. Adsorption of milk proteins onto the enzyme support can also sterically hinder the immobilized enzyme thus limiting its use. Woychik et al.[1] have suggested that the *K. lactis* enzyme would be most useful in the soluble form. However, this enzyme in an immobilized form may prove quite useful in hydrolyzing lactose in permeates from the ultrafiltration of sweet whey having a pH near 6. In contrast, the lactase with an acidic pH optimum from *A. niger* should prove quite suitable in an immobilized form for hydrolysis of lactose in acidic fluid products such as acid whey from the manufacture of cottage cheese. Additionally, the lower pH of the acid whey should minimize sanitary problems emanating from the growth of adventitious microorganisms.

3 Immobilization of Lactase

Immobilization of enzymes involves the physical or chemical localization upon an insoluble matrix or within a confined volume; the most common methods of immobilization are listed in Table 3[7]. Covalent binding of enzymes to insoluble supports provides a stable immobilized enzyme system but preparation of these products can be complex and require technical expertise. Expensive supports are often necessary which further increases the overall cost of the enzyme preparation. Additionally, a substantial excess

Table 3. Chemical and physical methods for immobilizing enzymes[a]

Chemical methods (covalent bond formation dependent)	Physical methods (noncovalent bond formation dependent)
Attachment to water-insoluble matrix	Adsorption onto water-insoluble matrix
Incorporation into growing polymer	Entrapment within water-insoluble gel matrix (lattice entrapment)
Intermolecular cross-linking with multi-functional, low-molecular-weight reagents	Entrapment within permanent and non-permanent semipermeable micro-capsules
	Containment within semipermeable membrane-dependent devices

[a] Adapted from original table by Zaborsky, O.R. In: K. Mosbach (ed.), Vol. 44. Methods in enzymology: Immobilized enzymes. New York: Academic Press 1976

of enzyme commonly is used in covalent attachment of enzymes to supports. Since enzymes are usually the most expensive components, methods for maximizing the utilization of enzymes must be developed.

Covalent attachment of lactase from *A. niger* to functionalized controlled-pore glass beads is the most frequently used technique for preparing immobilized lactases (IML)[8–21]. Other inorganic supports, such as silica[10] and titania[10, 11], have been employed but to a lesser extent. Of the organic supports, polysaccharides such as agarose[22], Sepharose[23–25], and aminoethyl cellulose[26] have been used with limited success. Nylon, in the form of nets[27] and tubing[28] has been utilized also for preparation of covalently bound lactases. Recently, a phenol-formaldehyde resin, Duolite[29], has been chemically derivatized for immobilization of neutral and acid lactases.

Entrapment techniques, involving cross-linked polymers, microcapsules, fibers, and membranes have been used to prepare lactase derivatives. Enzyme conjugates of widely different physical forms are feasible with the use of cross-linked polymers[30]. This has stimulated researchers to prepare polymerized beads[31–33], polymer-metal composite particles[34], and various gels such as acrylamide[35] and vinylpyrrolidone[36, 37] for immobilization of lactase. Unfortunately, these polymeric derivatives occasionally allow leakage of enzyme and effective immobilization often is dependent on a delicate balance of experimental factors[30] which minimizes their usage in large-scale applications.

Encapsulation can yield immobilized enzyme configurations with an extremely large surface area to volume ratio[30], making it an attractive method for immobilization of lactase if substrate and products permeate the microcapsules. Cellulose nitrate[38], nylon[39], and collodion[40] have been used for preparing microcapsules to entrap lactase. A distinct disadvantage of these preparations is the solubility and diffusion-dependent transfer of substrates and products through the capsule[38].

The relatively simple technique of immobilizing enzymes within or between supporting matrixes has prompted the investigation of ultrafiltration (UF) membranes[41], cellulose fibers[42], and membranous collagen[43, 44] as viable lactase supports. Initial data with collagen membranes suggest that the enzymic activity of entrapped lactase is limited by internal mass transfer effects[44]. Entrapment of enzymes in synthetic fibers is claimed to be well-suited for large-scale applications[45]. Lactase immobilized in cellulose acetate fibers is used in a successful commercial operation for hydrolyzing lactose in milk[46, 47].

Among the physical methods for immobilization, adsorption is probably the simplest and most economical approach to lactase insolubilization. Lactases have been adsorbed onto alumina[48–51], phenol-formaldehyde resins[3, 52, 53], nickel oxide[48–50], stainless steel[48–50], glass[20, 21], sand[20, 21], anion exchange resins[54, 55], cellulose powder[56], feather protein[57], chitin[58], and carbon[59] with various degrees of success. However, enzyme desorption can occur during continuous usage, particularly with a change in pH or ionic strength. This may be obviated if the enzyme is cross-linked by a difunctional reagent (such as glutaraldehyde) subsequent to its adsorption.

4 Stability of Immobilized Enzymes

Attaching enzymes to insoluble carriers enables their continuous use over extended periods of time. Consequently, the half-life of an IML preparation is a critical factor in determining the economic feasibility of a lactase reactor system. Controlling various reactor operating conditions to maximize half-life for the enzyme should be considered as a major economic factor[5]. The durability of a lactase reactor is affected by temperature, substrate flow rate, carrier material, immobilization method and substrate components. A common problem encountered with IML systems is enzyme inactivation by non-lactose components of fluid dairy products. This inactivation has been attributed primarily to the proteins in milk or whey and to various milk salts[9–11, 31, 52, 60, 61]. Inactivation by these milk components was illustrated by a comparison of stability of IML during treatment of whole acid whey vs. lactose[1]. The half-life of IML during treatment of whey at 50 °C was 8 d but was 100 d when 5% lactose was treated. Modifying surface characteristics of a support to minimize or eliminate adsorption of milk proteins would partially alleviate this problem[62]. Deproteinizing whey before treatment with IML as suggested by Coughlin and Charles[63] is an obvious practical solution but use of the product would be limited to a substitute sweetener.

Reactor temperature must be chosen carefully since it affects the extent of bacterial growth in the reactor and stability of the enzyme. The relative importance and effect of temperature, enzyme desorption and attrition of support and the interactive effects of these factors varies with different reactor systems. Consequently, it is impossible to make any general recommendations and each system has to be evaluated separately. Substantial decreases in stability at temperatures close to the "optimum" for IML have been reported in several studies[9, 11, 12, 52, 53]. Therefore, in terms of overall reactor efficiency it would be advantageous to operate at a temperature where no significant thermal inactivation of the enzyme would occur, but under conditions which allow only minimal bacterial growth. This parameter, of course, would depend on the particular reactor system employed and could be estimated by operational half-life studies. Enzyme leakage has been implicated also in decline of catalyst activity[36, 48, 61, 64, 65]. This can be partially avoided by proper choice of support and immobilization method. Again, the reactor system employed would also influence this decision. Catalyst decay can be attributed often to instability of the supporting matrix under operating conditions.

Attrition of support that has insufficient structural integrity can cause significant losses of enzymic activity of a reactor. Reagan et al.[26] found that the performance of a reactor decreased significantly during operation if support material disintegrated into smaller particles that were removed from (or which plug) the reactor. A loss of small particles from the reactor, although constituting a small weight percent loss of support, will result in a disproportionately large loss in enzymic activity since particles that are lost have a large surface to volume ratio.

5 Microbial Growth

The problems of unrestrained microbial growth during continuous use of IML could ultimately prevent the use of any potentially successful enzyme reactor system. In many food systems in which immobilized enzymes are utilized the food constituents, temperatures and pH values are ideal for proliferation of microorganisms[66]. Sources of microorganisms can be food components[8], processing equipment, and the enzyme preparation[67]. Several researchers have reported microbial growth concurrent with loss in lactase activity[8, 9, 19, 47]. Microorganisms affect the operational stability of the reactor in various ways[23]. Proteolytic exoenzymes may degrade the IML and have a deleterious effect on the food product. The support could also be attacked and degraded by microbial enzymes. In addition, plugging of the reactor (especially packed beds) may result from microbial proliferation.

Various approaches to the control of microbial contamination have been attempted. Harper et al.[8] studied effects of operating the enzyme column at pH and/or temperature values that would not be conducive to microbial growth. These researchers suggested that their *A. niger* lactase columns could be operated continuously for 48 h at pH 3.5 and at 50–60 °C without loss in activity.

Another approach to maximizing reactor efficiency is in the design of the reactor system. Mattiasson[23] suggested entrapping the enzyme in a "tight" support such as the cellulose acetate fibers used by Pastore et al.[46, 47, 68] to prevent destructive exoenzymes from attacking the catalyst. It may be advantageous to use inorganic or non-biodegradable organic supports. Also, use of stirred tank or fluidized bed reactors would minimize plugging of a reactor by bacterial contaminants or food particles[23]. Presently, use of antibacterial agents is the most common means of controlling microbial growth in IML reactors. Many sanitizers have been evaluated in fluid milk systems[3, 8, 10, 15, 46–48, 67] but various commercial quaternary ammonium compounds have been most effective[8, 46, 47, 67]. Other methods such as an iodophor treatment[48, 67] and a potassium sorbate treated feed solution (deoxygenated by CO_2 flushing)[3] have also shown promise in continuous flow enzymic systems.

A unique approach in sanitation is a self-sterilizing enzyme column. Lysozyme has been co-immobilized with lactase as a means of generating antibacterial activity in enzyme columns[23]. Preliminary investigations with this model system showed an enhanced resistance towards a medium infected with *Bacillus subtilis*. Further studies indicated that bacterial contamination was still a problem during long-term use of reactors. The co-immobilization of glucose oxidase and lactase has been studied as a means of improving the bacteriological quality of milk[69]. Glucose oxidase generates H_2O_2 which reacts with certain milk components to generate antibacterial substances. This system may eventually find applications in milk processing if residual H_2O_2 is eliminated.

6 Engineering Considerations

One of the most important considerations in designing an IML system is the choice of the reactor. Factors that determine the choice of reactor type are generally dependent on the kinetics of the reaction, operational requirements, frequency of catalyst replacement or regeneration, reactor costs and utilization[30]. Various types of IML reactors that have been used for process operations, listed in decreasing order of usage, are batch reactors, packed beds, recycle reactors, fluidized beds, and continuous stirred tank reactors (CSTR).

Based on the mode of charging/discharging, enzyme reactors may be generally classified as batch and continuous-flow reactors[45]. The batch reactor is simple, needs little ancillary equipment, and therefore is very suitable for small-scale experimental studies[30], which accounts for this reactor type receiving so much attention for use with immobilized lactases. The two main types of continuous reactor systems are the packed-bed and CSTR[70]. The packed-bed, proven to be more efficient than batch reactors for lactose hydrolysis[17, 18], requires high pressure drops for reasonable throughputs[49] and is easily plugged[36]. These problems increase in magnitude as the size of the catalyst particle decreases[49]. The CSTR systems are limited because they are mechanically complex, have high shear forces which lead to catalyst deterioration and efficiency decreases in the presence of product inhibitors of the enzyme[49], such as galactose. This tpye of reactor has found little application in IML systems[38].

A hybrid of the packed-bed and CSTR is the fluidized bed reactor. Coughlin et al.[50] discussed advantages of fluidized bed reactors, particularly for streams such as cheese whey which can contain particulate matter. Fluidized beds are capable of attaining levels of lactose hydrolysis at least as great as those obtained from comparable packed-beds. They operate more efficiently with smaller particles, which is a kinetically favorable situation. Also, fluidized beds are least attritive to support materials and are not subject to mechanical limitations associated with packed beds[49]. Consequently, they appear to have great potential as a reactor configuration for IML.

Another viable alternative for IML is the recycle reactor which can be combined with many of the basic reactor types discussed above, because a portion of the outflow is recycled and mixed with the inlet stream to the reactor[45]. Furthermore, it is possible to minimize bulk (external) mass transfer resistance since the reactor design inherently allows operation at high fluid velocities[45]. This system has been successfully used in an industrial plant, utilizing fiber-entrapped lactase from *K. lactis* for hydrolyzing lactose in milk[68].

7 Reactor Kinetics

Proper control of reactor kinetics often has a significant effect on efficiency of enzyme utilization. In general, kinetic parameters for lactases on solid supports can be affected by: 1) changes in the enzyme molecule, or in its immediate vicinity (an altered local microenvironment), that arises directly from the attachment of the enzyme molecule to the carrier[45], and 2) the rate processes of mass transfer[70]. Changes in the microenvironment usually cause a shift of pH and/or temperature profiles of the bound enzyme[44]. Numerous examples of these altered parameters for bound lactases have been reported[12, 29, 37, 44, 57, 58, 64].

Diffusional resistances (external and internal) are very important in immobilized lactase reactors. The slow diffusion (external mass transfer) of lactose between the bulk solution and the outer surface of the enzyme membrane or particle often results in a decreased reaction rate[44, 47, 65, 70]. The limitation was alleviated by increasing the flow rate of reactors thereby reducing the size of the Nernst layer[44, 47, 65]. Recycle reactors are well suited to minimize diffusional resistance. Diffusion (internal mass transfer) of lactose to the domain of the enzyme and diffusion of the products, especially the competitive inhibitor galactose, away from the internal domain of the lactase catalyst to the outer surface of a porous carrier may be an important rate limiting step of lactose hydrolysis[26, 44]. The severity of this limitation has been diminished by reducing particle size or membrane thickness of enzyme supports[26, 44, 65, 70]. A smaller particle also fluidizes better in a fluidized bed reactor[49] which increases efficiency of this reactor system.

8 Applications

Some advantages and disadvantages of immobilized enzymes as related to food processing are listed in Table 4. The decision to use immobilized enzymes in food processing must be predicated on a large number of considerations but the paucity of applications of immobilized enzymes in food systems makes evaluation very difficult. Those processes in which immobilized enzymes are being used currently on a commercial scale such as isomerization of glucose or resolution of L from D-amino acids are relatively simple. In

Table 4. Soluble vs. immobilized enzymes in food processing: advantages and disadvantages[a]

Disadvantages of soluble enzyme[b]

1. Large amount of enzyme can remain in product to participate in further (undesirable) reaction
2. Reuse of enzyme precluded
3. Extent of reaction limited by product inhibition
4. Precise control difficult
5. If necessary, enzymatic reaction generally stopped by heating, which can be detrimental to the food as well as add another processing step
6. Largely a batch process

Advantages of immobilized enzyme

1. Enzyme reusable
2. Reaction easily terminated by separating substrate from enzyme
3. More precise control
4. Less product inhibition
5. Greater pH and temperature stability
6. Can use enzymes presently unusable for various reasons
7. Potential operation over greater pH range by modifying charge characteristics of support
8. Continuous or batch use
9. Greater reactor design flexibility

Disadvantages of immobilized enzyme

1. Lower specific activity
2. Inactivation with continued operation
3. Cost of support and immobilization procedure
4. Need for sanitizing reactors

[b] Adapted from original table by Taylor et al. Coagulation of milk with immobilized proteases: a review. J. Milk Food Technol. *39*, 864 (1976).

[a] Advantages of soluble enzymes are not listed. The soluble enzyme is the presently used form and thus implicitly has advantages.

this regard, treatment of whey permeate from ultrafiltration systems with immobilized lactase may prove to be more attractive than hydrolysis of lactose in milk or untreated whey.

8.1 Commercial and Pilot Plant Operations

Presently, a small milk-processing plant in Italy is the only known commercial operation utilizing immobilized lactase technology[68]. Extensive pilot plant experiments led to this development[46, 64]. Lactose in milk is hydrolyzed in a recycle reactor utilizing spun cellulose acetate fibers containing entrapped lactase from *K. lactis*. The processed milk is claimed to retain its original organoleptic properties and exhibits only a slight sweeter taste.

Other pilot plant operations with cost estimates of scale-up have been described[3, 10, 63]. In all cases, the enzyme used was that from either *A. niger* or *K. lactis*. Coughlin et al. continuously processed acid whey on a pilot plant scale in the United States. Lactase from *A. niger* was adsorbed on alumina and cross-linked with glutaraldehyde and used in a fluidized-bed reactor capable of processing 1500 l of whey d^{-1}. The treated whey was also deproteinized and demineralized to increase acceptability of the final product as a natural sweetener in products such as ice cream, yogurt, eggnog and other dairy products. An estimated return on capital investments per year (based on 4.5×10^4 kg processed whey d^{-1}) in excess of 50% makes this system economically attractive[63].

Pitcher et al.[10] conducted scale-up experiments for continuous hydrolysis of deproteinized, demineralized acid whey using lactase from *A. niger* covalently bound to porous silica. The half-life of packed bed reactors was 62 d during treatment of acid whey. As of March 1975, the estimated overall cost to produce a 50% hydrolyzed-lactose product (concentrated to 50% solids) was approximately 18 to 22 c kg^{-1} of lactose, based on a projected plant capacity of 4.5×10^3 kg of processed lactose (in acid whey) d^{-1}. This compared favorably to the then current price of 37 c kg^{-1} (dry basis) for corn syrups. The availability of a commercial-scale IML plant in 1979 has been announced by Corning Glass Works, New York[71]. Deproteinized whey (optional demineralization) will be continuously treated in a packed-bed reactor followed by vacuum concentration of the hydrolyzed syrup. Estimated costs expressed as c kg^{-1} of hydrolyzed lactose are 9–13 c for demineralization by ion exchange, 9–11 c for 80% hydrolysis of lactose, and 7 c for concentration of syrup to 50–60% solids. It was assumed that no new building was required for a product throughput of 200 m^3 d^{-1}. A zero value was placed on the deproteinized whey which may be the permeate from an ultrafiltration system.

Okos et al.[3] recently investigated the feasibility of using immobilized lactase from *A. niger* adsorbed onto a phenol formaldehyde resin for commercial hydrolysis of acid whey. Their catalyst retained nearly full activity for over 120 d of continuous hydrolysis of lactose in acid whey and in pure lactose solutions. Based only on the cost of the enzyme and support, a conversion of 40% of the lactose could be achieved for 4 c kg^{-1} of lactose (1978 estimate). Higher degrees of lactose conversion dramatically increased the estimated cost of production.

In summary, the concept of industrial IML reactors has evolved into a reality. Extensive pilot plant work has led to the development of the first known industrial application of IML for the production of low-lactose skimmed milk. Processing of cheese whey with commercial IML reactors may be a profitable alternative to whey disposal problems. Subsequent improvements in IML technology will further enhance the desirability of utilizing IML reactors for processing of fluid dairy products.

The various IML systems discussed in this review are summarized in Table 5.

8.2 A Summary of Various Immobilized Lactase Systems

Table 5

Lactase	Immobilization	Reactor	Remarks	Ref.
Aspergillus niger	C-Glass	Packed bed	Retained substantial activity after storage at 55 °C and use in acid whey	14)
	C-Glass	Packed bed	Effectively sanitized by back washing column with 0.25% aqueous toluene	13)
	C-Glass	Batch	***Aspergillus niger*** had highest initial bound activity of various microbial preparations	15)
	C-Glass	Packed bed Batch	Packed bed more efficient than batch system	18)
	C-Glass	Batch	Temperature and pH optima unaffected by immobilization	17)
	C-Glass C-TiO_2	Batch	Compared commerical lactase preparations	11)
	C-Glass C-Silica C-TiO_2	Packed bed	Estimated overall cost of lactose hydrolysis (deproteinized, demineralized acid whey) was 18–22 c kg^{-1} vs 37 c kg^{-1} (dry basis) for corn syrups	10)
	C-Glass	Packed bed Batch	Packed bed more efficient than batch system	12)
	C-Glass	Packed bed	Exposure of IML (immobilized lactase) to acid whey resulted in partial, irreversible inhibition of activity	9)
			Analyzed effects of temperature of reactor operation on microbial growth and IML activity	8)

C-Glass	Batch	Co-immobilized glucose oxidase to generate H_2O_2 for the lactoperoxidase antibacterial system in milk	69)
C-Sepharose 4B	Packed bed	Co-immobilized lysozyme as a model study for self-sterilizing enzyme columns	23)
C-Duolite A 7	Packed bed Batch	Characterized *A. niger* IML and *K. fragilis* IML	29)
A*-Duloite	Packed bed	Demonstrated potential usefulness of Duolite resin for immobilization of lactase	53)
A*-Duolite S-30	Packed bed	Data indicated more efficient binding of enzyme as particle size of resin was reduced	52)
A*-Duolite	Fluidized bed	Estimated cost for lactose hydrolysis (in 5% lactose solution) was 4 c kg^{-1} lactose at 40% conversion	3)
A*-Chitin A*-Feather protein	Packed bed Batch	Feather protein retained approximately 8 times more activity than chitin on dry weight basis	58)
A*-Alumina NiO	Fluidized bed	Demonstrated a high binding efficiency (60–70%) with A*-Alumina preparation	51)
A-{Alumina, S. Steel, NiO} $TiCl_4$ treated		Fluidized bed with small particles very efficient method to treat acid whey	49)
		Estimated 132% annual return on investment utilizing A*-alumina catalyst for acid whey treatment	50)
E-Polymer-metal composite particles	Fluidized bed	Poor activity compared to alumina catalyst	34)
E-Polymerized gel using polyvinylpyrrolidone (PVP)	Packed bed	PVP method very simple and can be used in the food industry	36)
		Analyzed IML activity on various substrates	61)

Table 5 (continued)

Lactase	Immobilization	Reactor	Remarks	Ref.
Aspergillus niger	E-Membranous collagen	Recycle reactor (batch-spiral configuration)	Analyzed various sanitation techniques	67)
			Analyzed effects of metal ions and whey proteins	60)
			Diffusional effects are very important in IML systems	44)
	C-Glass and sand A-Glass and sand	Batch	Metal salt activation was inferior to glutaraldehyde treated silanized supports	20, 21)
	E-Cellulose fibers	Batch	Recovery of bound lactase activity was low (5–7%), but was doubled with glutaraldehyde treatment	42)
Kluyveromyces lactis	E-Comminuted collagen C-Glass	Packed bed	Slightly higher K_m for both bound enzymes but V_m were comparable for free and bound enzyme	19)
	E-Polyacrylamide beads (emulsion polymerization)	Batch	Immobilized 75% of the enzyme at optimum polymerization conditions	31)
	E-Cellulose triacetate fibers	Recycle reactor (batch tubular configuration)	Studied the kinetics of lactose hydrolysis in milk by fiber-entrapped enzymes	47)
			Results of long-term experiments on pilot plant utilizing fiber-entrapped lactase for reducing lactose in milk	46)
	C-Agarose	Batch Packed bed	Bound enzyme on "tether" to minimize steric hinderance	22)
	A-Carbon	Batch	Bound enzyme exhibited low level of initial activity and rapid loss of activity	59)
Escherichia coli	E-Cellulose triacetate fibers	Batch	Compared free and immobilized lactase from *E. coli* and yeast	64)

E-Cellulose nitrate membrane microcapsules	Packed bed and CSTR	Membrane permeability was limiting mass transfer resistance for overall reaction	38)
E-Nylon microcapsules by interfacial polymerization	Batch	Produced stable and active enzyme-microcapsules	39)
E-Collodion microcapsules by interfacial precipitation	Packed bed	Encapsulation procedure is milder than method of preparing nylon microcapsules	40)
E-Cross-linked 2-hydroxyethyl methacrylate (HEMA) gel	Recycle reactor (packed bed configuration)	Eliminated external diffusional effects by operating at high recirculation flow rates	65)
E-Spherical agar particles	Packed bed	Examined kinetics of immobilized whole-cell lactase under the influence of interparticle mass transfer	70)
C-Aminoethyl cellulose	Batch	Studied attrition of support material resulting from shear forces or abrasion	26)
C-Glass	Batch	Characterized lactase from *Escherichia coli* ATTCC-26 in both soluble and immobilized forms	16)
E-Membranes (collagen)	Continuous flow (spiral reactor configuration)	Studied the effects of structural modification on the enzyme-binding capacity of membranous collagen	43)
E-Polyacrylamide (gel) beads	Batch	Compared lactase activities of immobilized whole cells of *L. bulgaricus, E. coli,* and *K. lactis*	32)
C-Sepharose	Batch	Utilized immobilized lactase for production of lactose-reduced milk powder	24, 25)
A-Cellulose (powdered)	Batch	Demonstrated potential of hydrophobically modified supports for non-covalent adsorptive immobilization of lactose	56)
C-Nylon tubes	Continuous flow	Lactase co-immobilized with glucose oxidase for the automated determination of lactose	28)

Table 5 (continued)

Lactase	Immobilization	Reactor	Remarks	Ref.
Escherichia coli	E-between UF membranes	Continuous flow	Enzyme immobilized in a membrane sandwich reactor	41)
	C-Nylon nets	Batch	Bound enzyme exhibited the same specific activity as the soluble enzyme against p-nitrophenyl-galactosides	27)
Aspergillus oryzae	E-Polymerized gel using 2-hydroxyethyl acrylate (HEA) & dimethylacrylamide (DMAA)	Batch	HEA gel exhibited enzyme "leakage"; 30–56% of the activity entrapped in the DMAA gels	35)
	E-Polymerized gel using N-vinyl-pyrrolidone monomer (VP)	Batch	Lost more than 90% activity for VP-enzyme	37)
	E-Polyacrylamide beads (emulsion polymerization)	Batch	Presence of enzyme protective agents during polymerization procedure enhanced bound enzymic activity	33)
Curvularia inaequalis	A-DEAE Sephadex	Packed bed	Enzyme-dye derivative exhibited a 26.5 d half-life with 5% lactose	54)
	A-Anion exchange resins	Batch	Dye-modified enzyme was strongly adsorbed onto resins	55)

9 Nomenclature

A	Adsorption
A*	Adsorption with glutaraldehyde cross-linking
C	Covalent attachment
E	Entrapment
IML	Immobilized lactase
UF	Ultrafiltration

10 References

1. Woychik, J.H., Holsinger, V.H.: In: Enzymes in food and beverage processing. Ory, R.L., St. Angelo, A.J. (eds.), p. 67. Am. Chem. Soc., Washington, D. C. 1977
2. Vujick, I.F., Lin, A.Y., Nickerson, T.A.: J. Dairy Sci. *60*, 104 (1977)
3. Okos, E.S., Grulke, E.A., Syverson, A.: J. Food Sci. *43*, 566 (1978)
4. Ford, J.R.: Enzyme Technol. Digest *4*, 23 (1975)
5. Shukla, T.P.: CRC-Food Tech. *5*, 325 (1975)
6. Wondolowski, M.W.: In: Proc. Int. Congr. Biodeterioration. Sharpley, J.M., Kaplan, A.M. (eds.), p. 1033. Science Publ. Ltd. 1976
7. Zaborsky, O.R.: In: Methods in enzymology: immobilized enzymes. Mosbach, K. (ed.), Vol. 44, p. 317. New York: Academic Press 1976
8. Harper, W.J., Okos, E., Blaisdell, J.L.: In: Enzyme Eng. Pye, E., Kendall, Wingard, L.B., Jr. (eds.), p. 287. New York: Plenum Press 1974
9. Okos, E.S., Harper, W.J.: J. Food Sci. *39*, 88 (1974)
10. Pitcher, W.H., Jr., Ford, J.R., Weetall, H.H.: In: Methods in enzymology: immobilized enzymes. Mosbach, K. (ed.), Vol. 44, p. 782. New York: Academic Press 1976
11. Weetall, H.H., Havewala, N.B., Pitcher, W.H., Jr., Detai, C.C., Vann, W.P., Yaverbaum, S.: Biotech. Bioeng. *16*, 689 (1974)
12. Weetall, H.H., Havewala, N.B., Pitcher, W.H., Jr., Detai, C.C., Vann, W.P., Yaverbaum, S.: Biotech. Bioeng. *16*, 295 (1974)
13. Wierzbicki, L.E., Edwards, V.H., Kosikowski, F.V.: J. Food Sci. *39*, 374 (1974)
14. Wierzbicki, L.E., Edwards, V.H., Kosikowski, F.V.: Biotech. Bioeng. *16*, 397 (1974)
15. Wierzbicki, L.E., Edwards, V.H., Kosikowski, F.V.: J. Food Sci. *38*, 1070 (1973)
16. Wondolowski, M.W., Woychik, J.H.: Biotech. Bioeng. *16*, 1633 (1974)
17. Woychik, J.H., Wondolowski, M.V.: Biochem. Biophys. Acta *289*, 347 (1973)
18. Woychik, J.H., Wondolowski, M.V.: J. Milk Food Technol. *36*, 31 (1973)
19. Woychik, J.H., Wondolowski, M.V., Dahl, K.J.: In: Immobilized enzymes in food and microbial processes. Olsen, A.C., Cooney, C.L. (eds.), p. 41. New York: Plenum Press 1974
20. Byrne, M.J., Johnson, D.B.: Biochem. Soc. Trans. *2*, 496 (1974)
21. Thornton, D., Byrne, M.J., Flynn, A., Johnson, D.B.: Biochem. Soc. Trans. *2*, 1360 (1974)
22. Kilara, A., Shahani, K.M., Wagner, F.W.: Lebensm.-Wiss. U. Technol. *10*, 84 (1977)
23. Mattiasson, B.: Biotech. Bioeng. *19*, 777 (1977)
24. Portetelle, D., Thonart, P.: Lebensm.-Wiss. U. Technol. *8*, 278 (1975)
25. Portetelle, D., Thonart, P.: Lebensm.-Wiss. U. Technol. *8*, 274 (1975)
26. Reagan, D.L., Dunnill, P., Lilly, M.D.: Biotech. Bioeng. *16*, 333 (1974)
27. Faulstich, H., Schafer, A., Weckauf-Bloching, M.: F.E.B.S. Lett. *48*, 226 (1974)
28. Inman, D.J., Hornby, W.E.: Biochem. J. *137*, 25 (1973)
29. Watanake, K., Yokote, Y., Kimura, K., Samejima, H.: Agric. Biol. Chem. *41*, 553 (1977)

30. Zaborsky, O.R.: Immobilized enzymes. Cleveland, OH: CRC Publ. 1974
31. Dahlquist, A., Mattiasson, B., Mosbach, K.: Biotech. Bioeng. *15*, 395 (1973)
32. Ohmiya, K., Ohashi, H., Kobayashi, T., Shimizu, S.: Appl. Environ. Microbiol. *33*, 137 (1977)
33. Ohmiya, K., Terao, C., Shimizu, S., Kobayashi, T.: Agr. Biol. Chem. *39*, 491 (1975)
34. Charles, M., Coughlin, R.W., Tedman, R., Beard, K.W.: Biotech. Bioeng. *16*, 1549 (1974)
35. Maeda, H., Suzuki, H., Yamauchi, A., Sakimae, A.: Biotech. Bioeng. *17*, 119 (1975)
36. Maeda, H.: Biotech. Bioeng. *17*, 1571 (1975)
37. Maeda, H., Suzuki, H., Yamauchi, A., Sakimae, A.: Biotech. Bioeng. *16*, 1517 (1974)
38. Wadeak, D.T., Carlsonell, R.G.: Biotech. Bioeng. *17*, 1157 (1975)
39. Ostergaard, J.C.W., Martiny, S.C.: Biotech. Bioeng. *15*, 561 (1973)
40. Paine, M.A., Carlsonell, R.G.: Biotech. Bioeng. *17*, 617 (1975)
41. Vorselak, P., McCoy, B.J.: J. Food Sci. *40*, 431 (1975)
42. Linko, Y., Pohjola, L., Viskari, R., Linko, M.: F.E.B.S. Lett. *62*, 77 (1976)
43. Lin, P.M., Giacin, J.R., Leeder, J.G., Gilbert, S.C., Wang, S.S.: J. Food Sci. *41*, 1056 (1976)
44. Barndt, R.L., Wang, S.S., Leeder, J.G.: J. Food Sci. *41*, 494 (1976)
45. Vieth, W.R., Venkatasubramanian, K., Constantinides, A., Davidson, B.: In: Immobilized enzyme principles. Katchalski-Katzir, E., Goldstein, L. (eds.), Vol. 1, p. 222. New York: Academic Press 1976
46. Pastore, M., Morisi, F., Leali, L.: Milchwiss. *31*, 362 (1976)
47. Pastore, M., Morisi, F., Viglia, A.: J. Dairy Sci. *57*, 269 (1973)
48. Charles, M., Coughlin, R.W., Paruchuri, E.K., Allen, B.R., Hasselberger, F.X.: Adv. in Exp. Med. and Biol. *42*, 213 (1974)
49. Charles, M., Coughlin, R.W., Paruchuri, E.K., Allen, B.R., Hasselberger, F.X.: Biotech. Bioeng. *17*, 203 (1975)
50. Coughlin, R.W., Charles, M., Allen, B.R., Paruchuri, E.K., Hasselberger, F.X.: AIChE Sympos. Ser. I., 1974
51. Hasselberger, F.X., Allen, B.R., Paruchuri, E.K., Charles, M., Coughlin, R.W.: Biochem. Biophys. Res. Comm. *57*, 1054 (1974)
52. Hyrkas, K., Viskari, R., Linko, Y., Linko, M.: Milchwiss. *31*, 129 (1976)
53. Olson, A.C., Stanley, W.L.: J. Agric. Food Chem. *21*, 440 (1973)
54. Baeva, V.S., Kozlov, L.V., Antonov, V.K., Tikhomirova, A.S.: Biokhimiya *42*, 706 (1977)
55. Zolotova, V.S., Tikhomirova, A.S., Kozlov, L.V., Antonov, V.K.: Biokhimiya *41*, 1360 (1976)
56. Butler, L.G.: Arch. Biochem. Biophys. *171*, 645 (1975)
57. Stanley, W.L., Watters, G.G., Chan, B.G., Kelly, S.H.: Biotech. Bioeng. *18*, 1351 (1976)
58. Stanley, W.L., Watters, G.G., Chan, B., Mercer, J.M.: Biotech. Bioeng. *17*, 315 (1975)
59. Liu, C.C., Lahoda, E.J., Galasco, R.T., Wingard, L.B., Jr.: Biotech. Bioeng. *17*, 1695 (1975)
60. Jakubowski, J., Giacin, J.R., Kleyn, D.H., Gilbert, S.G., Leeder, J.G.: J. Food Sci. *40*, 467 (1975)
61. Giacin, J.R., Jakabowski, J., Leeder, J.G., Gilbert, S.G., Kleyn, D.H.: J. Food Sci. *39*, 751 (1974)
62. Cheryan, M., Van Wyk, P.J., Richardson, T., Olson, N.F.: Biotech. Bioeng. *18*, 273 (1976)
63. Coughlin, R.W., Charles, M.: In: Enzyme technology grantees-users conference. Pye, E.K. (ed.), p. 107. Philadelphia, PA: Univ. Penn. Press 1976
64. Morisi, F., Pastore, M., Viglia, A.: J. Dairy Sci. *56*, 1123 (1973)
65. Hinberg, I., Korus, R., O'Driscoll, K.F.: Biotech. Bioeng. *16*, 943 (1974)
66. Olson, N.F., Richardson, T.: J. Food Sci. *39*, 653 (1974)
67. Barndt, R.L., Leeder, J.G., Giacin, J.R., Kleyn, D.H.: J. Food Sci. *40*, 291 (1975)
68. Pastore, M., Morisi, F.: In: Methods in enzymology: immobilized enzymes. Mosbach, K. (ed.), Vol. 44, p. 822. New York: Academic Press 1976
69. Bjorck, L., Rosen, C.G.: Biotech. Bioeng. *18*, 1463 (1976)
70. Toda, K.: Biotech. Bioeng. *17*, 1729 (1975)
71. Corning Glass Works: a problem becomes an opportunity. IBS-10, Corning Glass Works, Corning, NY 1978

Immobilized Enzymes in Analytical Chemistry

Larry D. Bowers
Department of Laboratory Medicine and Pathology
University of Minnesota Medical School
Minneapolis, MN 55455, USA

Peter W. Carr
Department of Chemistry, University of Minnesota
Minneapolis, MN 55455, USA

The use of immobilized enzymes in analysis has increased tremendously in recent years. The various applications can be categorized by the manner in which the enzyme is used as: transducer bound enzymes; immobilized enzyme reactors; and miscellaneous batch and membrane applications. Enzyme electrodes, making use of either potentiometric or amperometric detection, are the best known example of the first class. Analyses based on reactors have been performed using immobilized enzymes either packed in a column as in liquid chromatography or bound to the inside walls of hollow tubes. In this review, we discuss the basic principles of operation of each of these classes with particular emphasis on optimization of the devices. In addition, some applications of immobilized enzymes in environmental pharmaceutical, and clinical chemistry are presented.

1 Introduction

The potential usefulness of enzymes in quantitative assays has been recognized for many years. There are a number of properties of these biocatalysts which make them particularly appealing in analyses. In particular the specificity of many enzymes allows their use for the determination of a substance in a complex matrix without prior separation. As a result, enzyme reactions have been used for over 130 years in the determination of substrates, activators, and inhibitors. Despite this long history, until recently the instability, scarcity, and expense of these materials and a general lack of familiarity with the principles of enzymology have discouraged many analysts from using them. In the past decade, advances in isolation and purification techniques have made over 325 of the 2000 known enzymes available in pure form at relatively low cost. Analysts in the pharmaceutical, food, clinical chemical, and environmental areas have responded with a voluminous number of enzyme based assays[1-3]. With the advent of a variety of immobilization techniques and a plethora of insoluble supports, it appears that the final obstacles have been removed from the use of enzymes as routine laboratory tools.

The analytical advantages of immobilized enzymes are summarized in Table 1. As shown, the keystone of the analytical benefits is the result of increased stability of the

Table 1. Analytical advantages of immobilized enzymes

1. Increased stability of enzyme
2. Repetitive use of a single batch of enzyme
3. Ability to readily remove the enzyme from a reaction mixture with no contamination
4. Long half-lives with predictable decay rates
5. Easy preparation of unstable or expensive reagents

enzyme in the immobilized state. Although this will be discussed further in a following section, it should be pointed out that increased stability does not always result after immobilization. The second advantage is the repetitive use of a single batch of enzyme for a large number of analyses. In some cases, the number of assays performed by a given amount of enzyme in the immobilized versus soluble state has been increased 5000 fold. Another advantage is the ease with which the enzyme can be recovered from the reaction mixture, particularly in the batch mode. When used in the transducer bound or reactor modes, the enzyme in essence becomes a part of the measuring device. As such, it can be calibrated along with the instrument in the same manner in which a pH electrode is calibrated in its system. The calibration process can in turn be used to assess the viability of the immobilized enzyme (vide infra). Finally, the use of these catalysts to prepare expensive or unstable reagents from inexpensive precursors is an area which has not, as yet, received much attention. This final advantage may become an important economic consideration, particularly in the area of clinical determinations. In the remainder of this chapter, we will review the analytical applications of immobilized enzymes with particular emphasis on the principles of operation and optimization of the enzyme analysis system. Because of the limitations of space, we will not attempt to review every application. A rather complete survey of this area has been published[4].

2 Immobilized Enzyme Technology

The immobilization of an enzyme or other active biochemical has been achieved by numerous specific means. To date, much of the literature in this field has been devoted to the characterization of a new carrier or immobilization reaction. This first stage of immobilized enzyme research is rapidly drawing to a close. It is now important to consider the economy, stability, ease of preparation, and usefulness in various physical forms, e.g., sheets, particles, etc., of the enzyme in the context of its potential use. In the remainder of this section, we will briefly consider immobilization techniques and the effect of these processes on the kinetics of the enzyme reaction.

2.1 Immobilization Methodology

Immobilization methods can be broken into two groups: physical methods, which include any process that does not involve the formation of covalent bonds, and chemical methods, which modify the protein through the formation of covalent bonds. Four methods have been used in the preparation of enzymes for analytical application: adsorption; covalent crosslinking of the enzyme to itself or a second protein; entrapment within a gel matrix; and covalent attachment to an insoluble carrier such as cellulose, polyacrylamide, or glass. A fifth immobilization technique, microencapsulation, has not been used in the area of analysis. As might be expected, the type of immobilization technique has a large affect on the analytical utility of the system. We will consider each of these methods below. For technical detail, the reader is referred to several compendia[5–9].

Adsorption. This process is relatively simple to perform chemically although initial optimization of the technique may be quite complex. Another advantage is the fact that in many cases almost all of the enzyme placed in contact with the support is recovered in the active state on the support. Adsorption has been widely used and offers one of the most economical approaches to obtaining a commercially useful preparation. Adsorption onto inorganic supports has been studied in detail by Zittle[10] and Messing[11]. More recently, Axèn, Porath and their coworkers have reported a hydrophobic adsorption technique[12]. A hydrophobic arm attached to the surface of the support interacts with hydrophobic regions on the enzyme to immobilize it. This method of immobilization would seem to have a number of advantages, particularly for lipophilic enzymes or enzymes with lipophilic substrates.

The major disadvantage of adsorption techniques is desorption. Different pH or ionic strength in the immobilization mixture as opposed to the analytical matrix may cause desorption and render the preparation of little analytical value. There have been reports of desorption caused by changes in the substrate concentration! In the case of hydrophobic adsorption, either salt or organic solvents may cause desorption. It should be noted, however, that a number of irreversibly adsorbed enzyme systems have been reported and are extremely useful[13, 14].

Gel Entrapment Techniques. The use of physical entrapment of an enzyme within the interstitial spaces of a polymeric matrix has been quite popular in analytical applications. Because of their size, enzymes within the polymeric lattice should not diffuse out during the polymerization process while appropriately sized substrate or product molecules can move in and out. The most popular matrices for gel entrapment include polyacrylamide[15, 16], silicon rubber[17], polyvinyl alcohol[18], starch[19], and silica gel[20]. There are a number of advantages to this approach including experimental simplicity, relatively mild reaction conditions, and the wide variety of forms in which it can be used.

There are several disadvantages. Control of numerous experimental factors during development, inactivation of the enzyme by radicals used in the polymerization process, and limitations on the substrate size are all troublesome in some applications, The most substantial problem, however, is leakage of protein from the polymer due to the heterogeneous poor size distribution. In all fairness, this leakage process may occur only in the early stages of use, after which the system is stable towards further leakage.

Protein Crosslinking. This technique has not found widespread use in an analytical setting. It is not easy to use since both intra- and intermolecular bonds can form and their distribution is difficult to control. Because intramolecular bonding is favored by low protein concentration, a second "carrier" protein such as albumin is often added to decrease this problem. The analytical uses of crosslinked proteins are limited to situations which require membranes. A number of "homogeneous" and "heterogeneous" bifunctional reagents used in this technique are shown in Fig. 1. More reagents will probably be developed for protein crosslinking due to the increased interest in enzyme-labeled antibodies and antigens.

Covalent Attachment. This final method of immobilization is the technique most frequently associated with the term "immobilized enzyme." There are essentially three steps in a covalent immobilization scheme: activation of the polymeric support; coupling of the enzyme; and removal of absorbed enzymes. Activation of the support is one of the most important aspects of the immobilization process. It can be a simple, one step process such as the use of cyanogen bromide with poly-hydroxyl compounds (especially polysaccharides) or a multistep process requiring a number of organic reactions as exemplified by the activation of nylon or the inorganic matrices. The expansion of covalent modification techniques has resulted in an impressive array of activating reactions, a detailed discussion of which is beyond the scope of this review[5–9]. Some common support materials are listed in Table 2. This is, in fact, an asset of the covalent attachment method. There are a number of factors which influence the characteristics of the enzyme conjugate prepared. The degree of activation has a significant influence on the recovery of enzyme activity. The greater the activation of the support, the greater the possibility of multiple bond formation to the enzyme possibly resulting in a restriction of molecular orientation with a concomitant loss of activity. It is in general easier

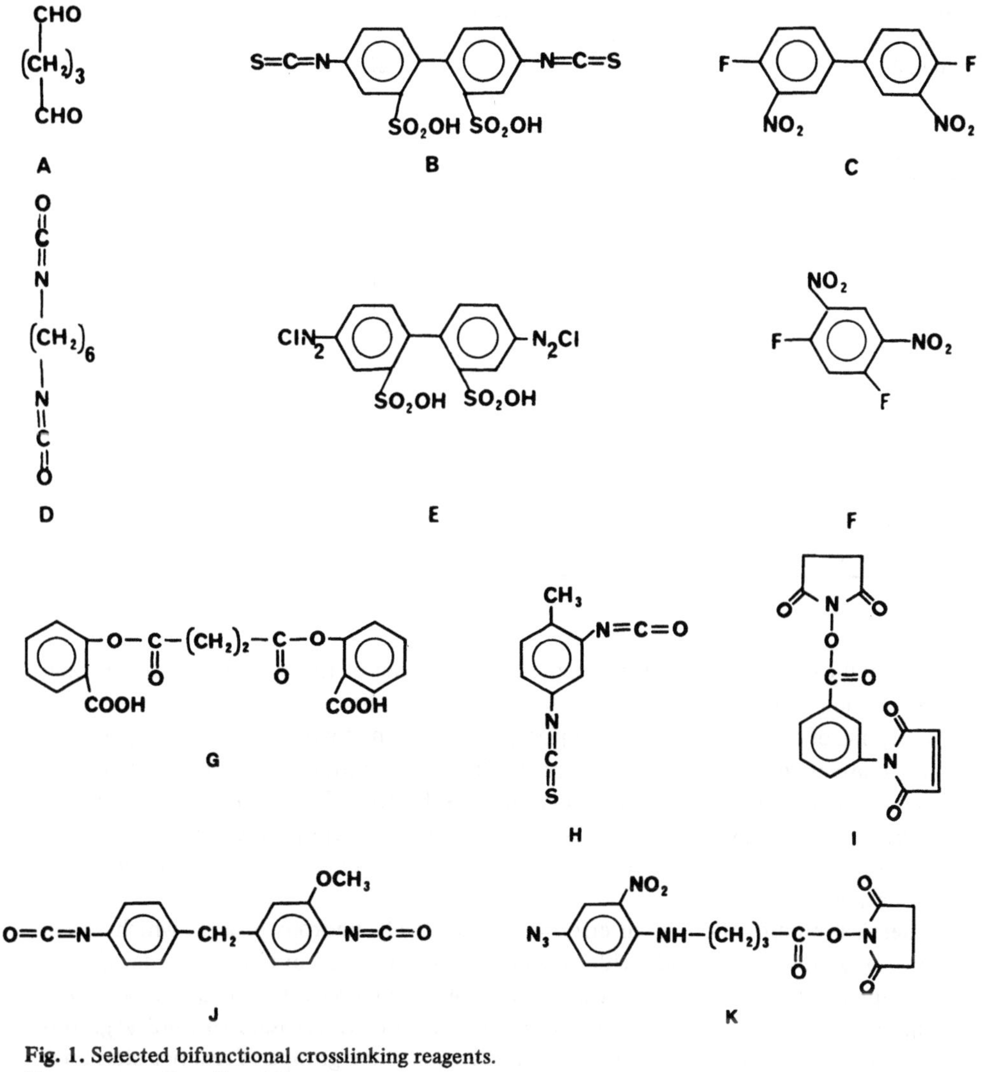

Fig. 1. Selected bifunctional crosslinking reagents.
Homogeneous Functional Groups:
A Glutardialdehyde[21]
B 4,4'-Diisothiocyanatobiphenyl-2,2'-disulfonic acid[24]
C 4,4'-Difluoro-3,3'-dinitrobiphenyl[23]
D Hexamethylenediisocyanate[25]
E bis-Diazobenzene[22]
F 1,5-Difluoro-2,4-dinitrobenzene[26]
G Succinyldisalicylate[30]
Heterogeneous Functional Groups:
H Toluene-2-isocyanato 4-isothiocyanate[27]
I Maleimidobenzoyl-N-hydroxysuccinimide ester[28]
J 3-Methoxydiphenylmethyl 4,4'-diisocyanate[27]
K N-(4-Azido-2 Nitrophenyl)-4-aminobutyrl N' oxysuccinimide[29]

Table 2. Support materials used for enzyme immobilization

Material	Ref.
Synthetic	
Acrylamide based polymers	31)
Maleic anhydride based polymers	32)
Methacrylic acid based polymers	33)
Polypeptides	34)
Styrene based polymers	35)
Acrylonitrile based polymers	36)
Nylon	37)
Dialdehyde starch-methylene dianiline	38)
Natural	
Agarose	39)
Dextran	39)
Glass	40, 41)
Titanium	41)
Collagen	42)
Magnetite	43)

to control the substitution by varying the concentration of the activating reagent rather than by manipulating reagents during the enzyme coupling step.

Coupling of the enzyme to the support generally consists of placing an enzyme solution in contact with the activated support. The concentration of enzyme used during the coupling step is an important consideration. In general, the use of 50–100 mg of enzyme per g of support will result in saturation of the coupling sites. In general, functional groups on the activated matrix have been capable of reaction with the system in aqueous media.

Chemical modification of an enzyme is a relatively rigorous process which should involve only those amino acid residues which are not involved in the active site. Groups which appear to be reactive towards functionalized matrices include α- and ϵ-amino groups, α-, β-, and γ-carboxyl groups, the phenolic portion of tyrosine, sulfhydryl groups, and the imidazole group of histidine. Mainly the first three groups of reactive sites have been used extensively in the immobilizations. The use of sulfhydryl groups is increasing in popularity due to the ease with which the coupling can be reversed. It should be apparent from the discussion of enzyme mechanisms that reaction of a residue involved in catalysis, such as serine in α-chymotrypsin, would seriously reduce the overall recovery of activity, the degree of the reduction being related to the overall number of reactive serine residues. Several techniques have been used to protect the active site by blocking it through combination of a competitive inhibitor or substrate with the enzyme, by combination of the enzyme with a reversible covalently linked inhibitor such as parachloromercuribenzoate with urease, by the addition of amino acids to the protein chain through which the covalent linkage can be made, or through the use of a zymogen which is the biological precursor of the enzyme of interest. It should be pointed out that this technique is not universally applicable nor does it always result in increased recovery of enzyme activity. The final step in the immobilization scheme, washing the support, is to remove protein which might otherwise contaminate the reaction mixture.

The major drawback to covalent coupling is the low recovery of activity on the support. Several new techniques, such as coupling to the carbohydrate moiety of a protein[44)] or the use of an imido ester intermediate[45)], appear to offer some relief from this problem.

2.2 Immobilized Enzyme Properties

The difference in the kinetic properties of a solution phase enzyme versus its immobilized counterpart has been recognized since the early days of immobilized enzyme research. The discrepancies are the result of three processes: rate limiting mass transfer; steric orientation; and microenvironmental effects. The limitations imposed by slow mass transfer of substrate to or product from the enzyme site in the matrix seriously affect the observed kinetics. Since mass transfer is a first order process with respect to substrate concentration, the expected, and observed, effect would be to extend the apparent first order kinetic region. Since this is the region of analytical interest, mass transfer can increase the linear dynamic range, improve precision, and decrease the temperature dependence in an analytical system. The effects of internal and external mass transfer on immobilized enzyme kinetics are discussed further in a later chapter and will not be considered here.

Second, one must consider the steric orientation of the enzyme. The enzyme may be coupled in such a way that the active site is partially or totally blocked or in an inactive conformation such that catalytic action is reduced in some manner. It is possible that these orientations contribute to the kinetic constant of the system as a whole although this has not been studied in depth.

The microenvironment of the enzyme, as dictated by the support, has a large affect on the kinetic parameters. There are two discernable cases in which this effect might be observed. The first of these might best be envisioned as an interaction of the charged portions of the support with the enzyme in such a way that the basic rate of the reaction is altered. This is a rather new area of research in which the rate constants of the forward and reverse reactions are determined and compared with the data from soluble enzymes. An example of this interaction between support and active site can be found in the coupling of chymotrypsin to ethylenemaleic acid coplymers through polyornithyl and polyglutamyl side chains[46)].

The second microenvironment effect is partitioning of a kinetic effector such as hydrogen ion, substrate, product, inhibitor or activator between the bulk of solution and the support. Because of the partitioning, the enzyme experiences a milieu different from the bulk of solution and thus different from the measured concentration. Thus for a negatively charged support, the local hydrogen ion concentration is higher than the solution concentration. As a result, one would expect a shift in the pH optimum to a more alkaline pH to compensate for the local pH. The opposite effect would be expected for a positively charged support. Goldstein and Katchalski[47] have studied this effect and a sample of their results are shown in Fig. 2. Similar results are observed for substrates where the observed K_m for an enzyme system with an oppositely charged substrate and matrix was an order of magnitude lower than the solution phase K_m[47]. Recently hydrophobic interactions have been shown to cause similar effects[48].

An additional change in the behavior of the enzyme, increased stability, was alluded to earlier. In many cases, the rate of diminution of the enzyme activity is decreased after immobilization. It is important to recognize that three types of stability can be measured: thermal stability, storage stability, and operational stability. Thermal stability is the ability of the enzyme to function at elevated temperatures which would normally result in denaturation of the protein. Storage stability is related to the loss of activity experienced when stored under some specified set of conditions. The ability of the enzyme to withstand the chemical and mechanical rigors of repeated use, e.g., in a continuous flow analyzer, determines the operational stability of the immobilized enzyme prepara-

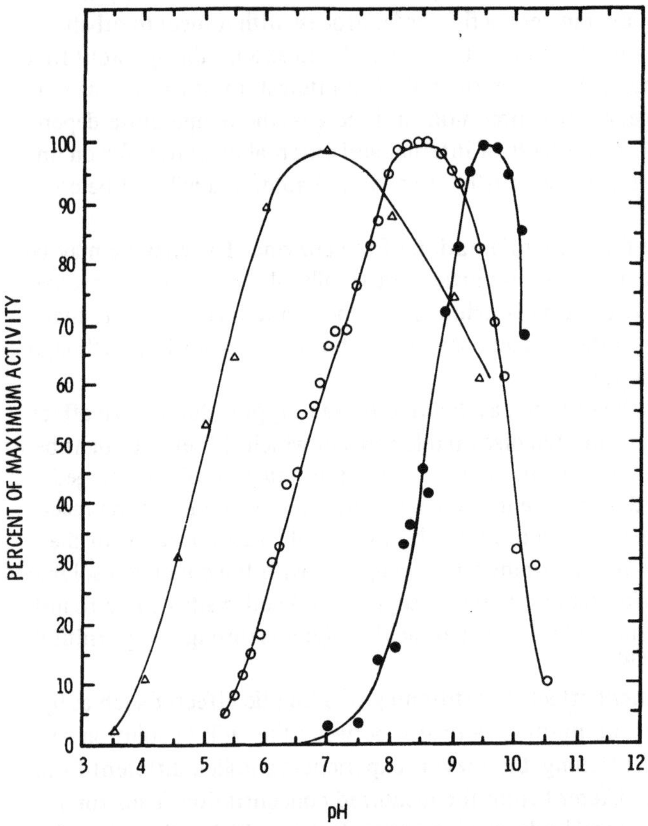

Fig. 2. Effect of the support matrix on the pH characteristics of an immobilized enzyme. The curves represent chymotrypsin (○), a polyanionic (ethylenemaleic acid) copolymer of chymotrypsin (●), and a polycationic (polyornithyl) derivative of chymotrypsin (△)

tion. These parameters are, of course, interrelated. The operational stability, for example, will be dependent on the thermal stability, particularly at elevated temperatures. Other factors affecting the operational stability are carrier durability, concentrations of inhibitors in the analytical system or samples, and features of the sample such as particulate content. As analysts, we are interested in the operational longevity of an immobilized enzyme. Unfortunately, at this time none of these stability factors are predictable.

An example of the unpredictability of the situation is shown in Table 3. This table, taken from a study by Weetall[49)], illustrates the variety of storage stabilities exhibited by glucose oxidase when various immobilization reactions and supports are used. For a number of supports, immobilization of the enzyme results in an increase in stability. It is interesting to note that the cellulose derivative not only has relatively poor apparent specific activity, but also is less stable in storage than the lypholyzed enzyme. The importance of the linking reaction is also illustrated by comparing the controlled pore glass (CPG) preparations. This preferential stability of linking reactions is not restricted to

Table 3. Stability of immobilized glucose oxidase derivatives[a]

Derivative	Original activity[b] (mg g^{-1})	Conditions of storage (°C)	Storage time (d)	Percentage activity
Alumina-CPG-isothiocyanate	6.0	5	68	100
		5, in water	68	100
		23	68	100
		23, in water	68	100
Hydroxyapatite isothiocyanato	10.7	5	68	100
		5, in water	68	65
		23	68	100
		23, in water	68	100
CPG-isothiocyanate	11.8	5	68	81
		5, in water	68	73
		23	68	73
		23, in water	68	77
CPG-azo	10.1	5	68	100
		5, in water	68	100
		23	68	100
		23, in water	68	100
Cellulose-azo	6.1	5	68	69
		5, in water	68	16
		23	68	57
		23, in water	68	9
Soluble enzyme		5	60	100
		23	60	75

[a] Data taken from Ref.[49)]

[b] Activity per gram of support material was calculated assuming that the immobilized enzyme retains the same specific activity as the soluble enzyme

storage stability. The resistance of lactate dehydrogenase to denaturation by urea solutions is significantly greater for the enzyme linked to alkylamine glass by glutaraldehyde than that bound by a diazo derivative[50].

2.3 Immobilization Rationale

From the above sections and from the other chapters of this book, it should be obvious to the reader that there is no "best" way to immobilize an enzyme. Immobilization itself is a somewhat empirical science. There are, however, a number of factors which might systematically limit the choice of an immobilization scheme.

As mentioned above, covalent bonding requires chemical modification of the enzyme. The reactive groups on the protein for the various immobilization reactions are known. It is imperative that amino acids not present in the active site if known, be chosen for modification. Thus, it is important to know the enzyme with which one is working. Many random, unsuccessful approaches could be avoided by a careful consideration of the enzyme involved.

The proliferation of support media which have appeared in the literature have been the result of a search for the perfect matrix. The attributes of a good carrier have been reviewed in this series by Messing[51]. These considerations also apply to the support used in an analytical system. The choice of a support is quite dependent on the system in which it is used. An enzyme electrode requires quite different support properties than does a packed bed reactor. The capacity, mechanical and chemical stability, expense, availability, ease of preparation, and flow characteristics should all be considered in the application. Although these factors do not always clearly indicate one "perfect" support, they do allow a logical selection. The above discussion and the remainder of this volume should serve as a useful guide in a rational approach to immobilization.

3 Transducer Bound Enzymes

The first type of application can be summarized by the term "transducer bound enzymes." The main features of such a device are a thin layer of immobilized enzyme (100–200 microns) and some sort of transducer capable of detecting a physical or chemical change in its immediate surroundings. The enzyme layer is positioned and held in close contact with the transducer. In some cases, a thin membrane might be placed between the enzyme layer and the transducer surface to improve the specificity of the sensor, e.g., the Clark O_2 electrode. A second semi-permeable membrane may be placed over the outside of the enzyme layer to secure the enzyme layer in place. When placed in solution the substrate diffuses across the Nernst layer and into the enzyme membrane where it undergoes conversion to product. The sensor system measures the concentration of product or co-reactant present at the enzyme membrane-transducer interface.

The best known example of a transducer bound enzyme probe are the so-called "enzyme electrodes." A schematic representation of this device is shown in Fig. 3. As mentioned above, when the electrode system is immersed into the analyte solution, conversion of substrate to product takes place in the trapped enzyme layer. Because the reaction takes place only in the enzyme layer, a concentration gradient of both substrate and product will be generated by the reaction. After some interval of time (30 s–10 min), a steady state condition will occur in which the rate of supply and the rate of consumption of substrate will be equal. At this point, a measurement can be made. Both the steady-state potential or current and the rate of change of the potential or current have been shown to be related to the analyte concentration and thus analytically useful.

Potentiometric electrodes are generally chosen so as to detect the product of the enzyme reaction. In particular, pH electrodes and ammonium ion selective electrodes have been used for hydrolytic enzyme based electrodes. Amperometric transducers, in contradistinction, measure the flux of the electroactive species they are set up to detect. They have been used most widely to determine the decrease in oxygen tension or the increase of product (H_2O_2) in enzymatic oxidation-reduction reactions.

The interest in enzyme electrodes has been phenomenal as illustrated by the appearance of over 90 application papers in the last decade, most in the last five years. A number of the analytical advantages which have lead to this interest are shown in Table 4 along with a few limitations. Although these systematic advantages are substantial, several other advantages accrue which defy simple compartmentalization and require a brief discussion. As with all immobilized enzyme applications, the enzyme electrode may be reused, as many as 10,000 times for some electrode systems. Second, if the enzyme can be immobilized or trapped with a high active concentration per unit

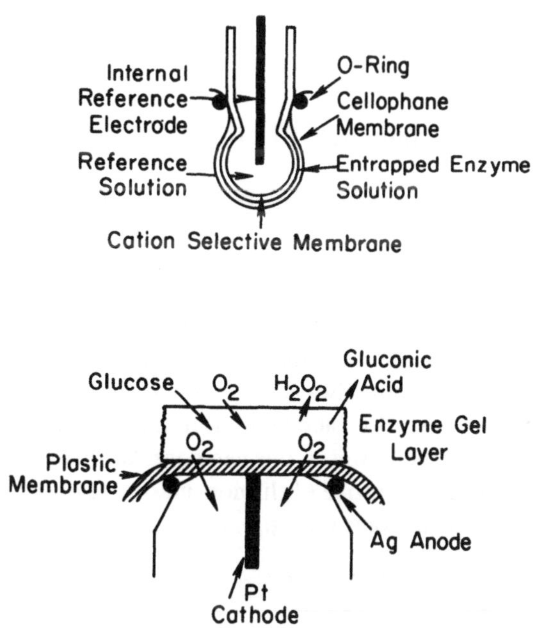

Fig. 3. Schematic representation of an enzyme electrode. *A* Potentiometric sensor; *B* amperometric sensor

Table 4. Advantages and limitations of enzyme electrodes

Simple, easy to perform analysis
Minimal sample preparation
Inexpensive, easy to calibrate
Good sensitivity
Response time (up to 10 min)
Limited to electroactive systems
Small sample volume possible

weight of support, the enzyme electrode response will be independent of the amount of enzyme present. This is never true in a kinetic assay with a soluble enzyme. A third advantage arises from the independence of the system response and enzyme reaction rate when a vast kinetic excess of enzyme is present: the response of the assay system to variables which influence the rate of the enzyme reaction, e.g., pH, inhibitors, activators, temperature, etc., is diminished. Thus the immobilized enzyme electrode systems may be more accurate and precise than soluble enzyme-based assays.

In the remainder of this section, we will discuss the principle of operation of both potentiometric and aperometric enzyme electrodes. In particular, the parameters which can be used to optimize enzyme electrode behavior will be presented within the framework of a simple steady state model. Various applications will then be discussed. Although we will consider only electrochemical transducers in this section, it should be recognized that there is no a priori reason why other transducers could not be used in a similar mode. In point of fact, thermistors have been used to measure the temperature change in a transducer bound enzyme membrane[52, 53)]. Recent reports indicate that chemiluminescence produced by an enzyme fixed at the end of a fiber optic light pipe may be analytically useful.

3.1 Potentiometric Electrodes

The sequence of events necessary for the operation of an immobilized enzyme probe are summarized in Fig. 3. As indicated, the immobilized enzyme membrane is in close contact with both the active surface of the electrode as well as the analyte solution. Ideally, the behavior of the system should be determined only by the chemical reaction and by diffusion through the enzyme layer. This requires that the primary transducer have a rapid response time, the membranes be ultrathin, and the bulk solution well stirred. Clearly this is not always the case and factors such as membrane thickness and permeability, convection, and non-linear kinetics tend to complicate the picture.

In general, however, one can consider three potential rate determining steps which will control the probe's behavior: mass transport from the bulk of solution to the outer membrane at point L′, e.g.,

$$S_{\text{bulk}} \xrightarrow[\text{transfer}]{\text{external mass}} S_{\text{at } L'} \; ; \tag{1}$$

diffusion through the outer membrane to point L, e.g.,

$$S_{at\ L'} \xrightarrow[\text{diffusion}]{\text{membrane}} S_{at\ L}\ ; \tag{2}$$

and chemical reaction in the enzyme layer. The latter process will, in simplest terms, be described by Michaelis-Menten kinetics

$$E + S \underset{k_{-1}}{\overset{k_1}{\rightleftharpoons}} ES \xrightarrow{k_2} E + P \tag{3}$$

the product formed in the reaction will diffuse back into the solution.

Racine and Mindt have considered the behavior of a potentiometric electrode limited by membrane diffusion and no substrate gradient present in the enzyme layer[54]. Under these conditions, the rate of reaction, $\frac{dN_s}{\bar{V}\,dt}$, will be determined by the rate of substrate entering the enzyme layer and given by the relation

$$\frac{dN_s}{dt} = \phi_s\,([S]_{L'} - [S]_L) \tag{4}$$

where N_s is the number of moles of substrate and ϕ_s is the membrane permeability. The permeability is related to its area, its thickness, and the ability of the substrate to pass through it. At steady-state, the rates of mass transfer and chemical reaction are equal and the equations can be solved simultaneously. In the limit of zero order kinetics ($K_m \ll [S]$), their result is

$$[P] = \frac{k_2[E]\bar{V}}{\phi_s} \tag{5}$$

whereas for a first-order reaction ($K_m \gg [S]$),

$$[P] = \frac{k_2[E]\bar{V}/K_m\phi_s}{1 + (k_2[E]\bar{V}/K_m\phi_s)}\,[S]_{bulk}\ . \tag{6}$$

Several conclusions can be drawn from these relations. First, as shown in Fig. 4, the response of an electrode sensing the product concentration is significantly affected by the Michaelis constant of the enzyme. Second, the sensitivity and linearity of the electrode are controlled by the relative ratio of the rates of reaction and mass transfer. At high enzyme concentrations ($k_2[E]\bar{V} \gg K_m\phi_s$), the product concentration will equal the bulk substrate concentration and maximum sensitivity will be obtained. A very important conclusion from these equations is that linearity may be obtained *at concentrations which exceed the K_m* when the rate of reaction is controlled by mass transfer. Mass transfer control also results in relative independence from other factors which alter en-

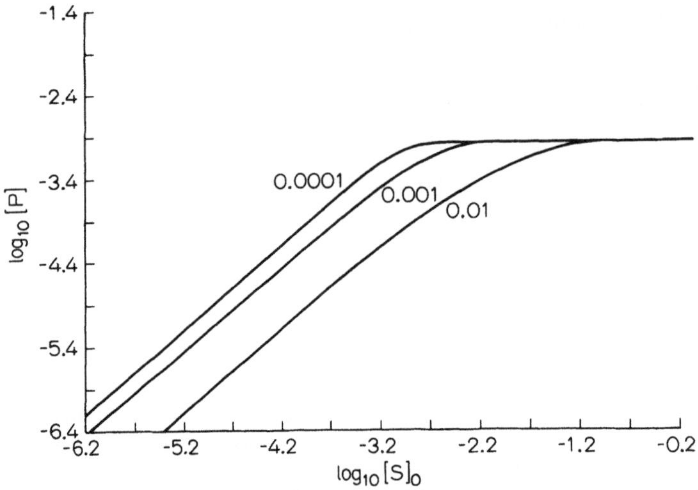

Fig. 4. Simulated response curves for a potentiometric enzyme electrode. Numbers on curves represent the K_m (mol l^{-1}) used in the calculation $\left(\frac{k_2\,[E]\bar{V}}{\phi_s} = 10^{-3}\right)$

zyme activity. It should be noted that decreasing the membrane permeability will also result in mass transfer limitations but at the expense of increased response times. In contradistinction to the preceding conclusions, slow reaction rates will result in loss of product by diffusion and a decrease in sensitivity.

If one ignores the effects of external and membrane diffusion and partitioning, it has been shown that the response of the electrode is a function of diffusion in the enzyme layer and the rate of reaction[18, 55]. Thus,

$$\frac{\partial\,[P]}{\partial t} = D_p\,\frac{\partial^2\,[P]}{\partial x^2} + \frac{k_2'\,[S]}{K_m + [S]}\,, \tag{7}$$

$$\frac{\partial\,[S]}{\partial} = D_s\,\frac{\partial^2\,[S]}{\partial x^2} - \frac{k_2'\,[S]}{K_m + [S]}\,. \tag{8}$$

Both the steady state and transient response can be obtained from these equations depending on the nature of the applied boundary conditions. If we assume that at the outer membrane the substrate concentration is the same as the bulk and no product is present, that the flux of the species at the electrode is zero, and that upon initiation of the experiment there is no substrate or product present in the enzyme membrane, the above equations can again be solved for the product concentration at the electrode surface yielding for the zero order region ($K_m \ll [S]$)

$$[P]_{x=0} = \frac{k_2\,[E]X^2}{2\,Dp} \tag{9}$$

and for the first order region ($K_m \gg [S]$),

$$[P]_{x=0} = \frac{Ds}{Dp}\,[S]_{bulk}\left(1 - \operatorname{sech}\frac{k_2\,[E]X^2}{K_m D_s}\right). \tag{10}$$

There are a number of interesting conclusions to be drawn at this point. First, from Eq. (9), at high substrate concentration the electrode potential will be independent of $[S]_{bulk}$. A second result of Eq. (9) is that one would expect a shift in the potential observed with changes in the membrane enzyme concentration or membrane thickness. This second prediction has not been consistantly observed[56–59]. One of the most important results is that linearity should not be expected at substrate concentrations in excess of K_m. Digital simulation studies[60] indicate that even when there is a substantial concentration gradient across the enzyme layer, i.e., $[S]_{x\to 0} \ll [S]_{bulk}$, the product concentration at the sensor surface is not linearly related to the bulk substrate concentration when $[S]_{bulk} \geq 0.2\,K_m$. This is the exact opposite of the behavior observed for a membrane diffusion limited electrode system. This area of electrode research requires more detailed experimental investigation to facilitate future development of optimal electrode systems.

In contrast to the zero order kinetic region, the first order region for internal diffusion is quite similar to that predicted for the external diffusion controlled system. The most important result of Eq. (10) is the linear relationship between the electrode response and the substrate concentration. Once again the enzyme loading factor

$$\frac{k_2\,[E]X^2}{K_m D_s}$$

is the most important variable in determing the electrode behavior. From Fig. 5, the relationship between the normalized electrode response factor and the loading factor can be appreciated. Obviously, maximum sensitivity and independence from kinetic effectors are obtained at large loading factors.

There are a number of additional factors which must be considered to evaluate the performance of an enzyme electrode system. These include the limit of detection, the response time, and the wash out time of the electrode. Wash out time is required to remove the product which was generated in the enzyme layer from the previous analysis. The wash time is generally regarded as the time required to re-establish the baseline potential. This regeneration time is dependent on the thickness of the enzyme layer, the enzyme used and the transducer. The wash out time for a urea electrode based on the air gap electrode, for example, is only 20 s; whereas that based on a glass pH electrode is on the order of 10 min[61, 62]. An automatic electrode washer has been reported[63].

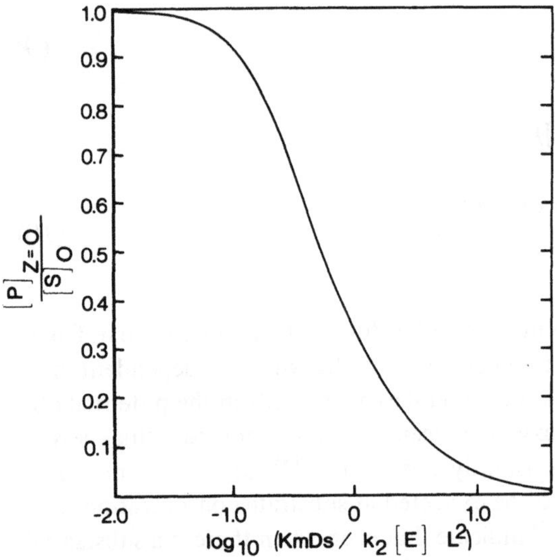

Fig. 5. Normalized electrode surface product concentration as a function of the enzyme loading factor. Results are limited to a first order reaction with internal diffusion control

The *response time* of an enzyme electrode is a function of a number of variables as shown in Table 5. Mascini and Liberti[64)] have studied the effects of external stirring rate, amount of enzyme, substrate concentration, pH, and thickness of the outer membrane on the steady-state response time of an amygdalin electrode. They found that rapid stirring, high substrate concentration, a pH near the optimum for the enzyme reaction, and an outer membrane thickness of less than 25 microns produced the most rapid response. Recently, the transient response characteristics of an enzyme layer limited electrode were evaluated theoretically[65)]. It was predicted from this work that in general the response time in the zero order region will be faster than that obtained in the first order region. This phenomenon has been observed experimentally. It is interesting, however, that at high enzyme loadings

$$\frac{k_2\,[E]\,X^2}{DK_m} \gg 1$$

the time constants for the two kinetic regions became similar; the result of diffusion control. A more detailed discussion is beyond the scope of this review, and the reader is referred to the literature[4, 66)].

The limit of detection of a potentiometric enzyme electrode is determined by the selectivity of the sensor and the enzyme loading as noted earlier. The effect of nonselectivity can be described by the relation

$$E = \text{constant} + \frac{RT}{zF}\ln\left([P]_{x=0} + \Sigma\, K_I[I]\right), \qquad (11)$$

where K_I is the selectivity coefficient for interferant I. The limit of detection will then be determined by the amount of interferant present.

Table 5. Factors affecting electrode response time

Physical factors
Stirring rate
Outer membrane thickness and permeability
Enzyme membrane thickness
Chemical factors
Enzyme loading
Substrate concentration
pH, temperature
Instrumental factors
Sensor response time
pH effects on probe

As mentioned earlier, a large number of enzyme electrodes have been developed. A cross section of the field is presented in Table 6 along with a few experimental details. More complete listings may be found in the literature[4, 66]. It has been our intention here to present the operational principles of these devices.

3.2 Amperometric Electrodes

Most of the preceding discussion pertains only to potentiometric electrodes since an amperometric electrode, by the measurement process, influences the amount of electroactive species at its surface. None the less, many analogies do exist between the two systems. The basic difference is that while the flux of electroactive species is still zero, the amperometric electrode will exhibit maximum sensitivity when all of the product is oxidized or reduced ($[P]_{x=0} = 0$). Once again, the flux of product to the electrode can be limited by external mass transfer or diffusion in the immobilized enzyme layer. Using the same basic assumptions as for the potentiometric probe, the relationship between the current and the analyte concentration can be obtained for zero order conditions where

$$i = nF\,k_2\,[E]\,\overline{V} \tag{12}$$

as well as for first order conditions where

$$i = \frac{nF\,k_2\,[E]\,\overline{V}}{K_m + \dfrac{k_2\,[E]\,\overline{V}}{\phi_s}}\,[S]_{bulk} \tag{13}$$

and n and F have their usual meaning[54]. It should be apparent that either term in the denominator of Eq. (13) can predominate depending on whether chemical reaction or

Table 6. Potentiometric enzyme probes

Substrate	Enzyme	Electrode	Response time	Linearity	Stability	Ref.
Urea	Urease (E.C. 3.5.1.5)	Beckman univalent cation	25–60 s for 98%	0.02–8 mg l^{-1}	> 14 d	58, 63, 69)
			60–90 s	10^{-4} – 10^{-2} M	> 21 d	70)
		Instrumentation laboratories CO_2 electrode	1–6 min	10^{-4} – 10^{-2} M	> 3 d	71)
		Nonactin in silicon rubber	60–80 s	3 x 10^{-5} – 3 x 10^{-3} M	> 7 d	72)
		Air gap	1 min @ > 10^{-2} M 4 min @ < 10^{-3} M	10^{-4} – 2 x 10^{-2} M	> 21 d (300 assays)	61)
		NH_3 gas sensor	1.5 – 2 min @ 10^{-2} M pH 9	5 x 10^{-4} – 5 x 10^{-2} M	~ 20 d	73)
		pH glass	–	5 x 10^{-5} – 5 x 10^{-3} M	> 14 d	62)
Glucose	Glucose oxidase (E.C. 1.1.3.4)	Iodide ion selective				74)
		pH electrode	–	10^{-3} – 10^{-1} M	> 14 d	62)
Amygdalin	β-Glucosidase AgI type	Heterogeneous	5 min	10^{-4} – 10^{-1} M	~ 7 d	64)
		CN^- solid state electrode	< 1 min	5 x 10^{-7} – 5 x 10^{-3} M	> 14 d	75)

Penicillin	Penicillinase (E.C. 3.5.2.6)	pH glass	> 2 min	$10^{-3} - 10^{-2}$	> 7 d	62)
		pH glass				58)
Creatinine	Creatininase (E.C. 3.5.4.2.1)	NH_3 gas sensor	6–10 min @ < 5 x 10^{-3} M	7 x $10^{-5} - 10^{-2}$ M	~ 4	76)
			2–5 min @ > 5 x 10^{-3} M			
L-amino acids	L-amino acid oxidase	Beckman univalent cation	1–2 min	$10^{-4} - 10^{-3}$ M	14 d	77)
Uric acid	Uric acid oxidase	CO_2 gas sensor	5–15 min	2 x 10^{-4} – 3 x 10^{-3} M	~ 10 d	78)

mass transfer are rate limiting. Thus, as shown for the potentiometric system, under external diffusion control linearity for the amperometric system may exceed the Michaelis constant.

A similar analogy can be drawn for limitations resulting from simultaneous reaction and diffusion in the enzyme layer. Mell and Maloy have carried out extensive experimental and theoretical simultions of amperometric electrodes under these conditions[67, 68]. As might be expected, the equations of best fit were dependent on the enzyme loading factor. At low loading factors $\frac{k_2 [E] X^2}{DK_m} \leqslant 1$, the relation

$$i = \frac{nF\bar{V}k_2 [E] [S]_{bulk}}{2 K_m + [S]_{bulk}} \tag{14}$$

fit the data. At high loadings $\frac{k_2 [E] X^2}{DK_m} > 10$, the simulated data were best described by the equation

$$i = \frac{nF\bar{V}k_2 [E] [S]_{bulk}}{\frac{k_2 [E] X^2}{D} + 2 [S]_{bulk}} . \tag{15}$$

When the amount of enzyme in the membrane is low, the linearity of the electrode is limited to the region where $[S]_{bulk} \ll K_m$. When mass transfer is the controlling step, it can be shown from Eq. (15) that

$$i = \frac{nFAD [S]_{bulk}}{X} \tag{16}$$

when $[S]_{bulk} \ll \frac{k_2 [E] X^2}{2 D}$. Analytically, this work presents three important results: 1) the current at high loading factors will be a linear function of the substrate concentration at concentrations greater than K_m; 2) the current at low substrate concentration will increase as membrane thickness decreases if mass transfer is rate limiting; and 3) the time to steady-state can be estimated from the relation

$$t \leqslant 1.5 X^2/D . \tag{17}$$

The performance characteristics of amperometric enzyme electrodes are also an important consideration in a potential application of the device. As with the potentiometric probe, the response time is influenced by the stirring rate and the substrate concentration. Wash out time is usually relatively short for the amperometric probe because the electrode can consume the active species. The limit of detection is determined by three factors: the magnitude of other electroactive species (background); the sensitivity of the technique; and the reproducibility of these factors. The background is difficult to predetermine and thus is an empirical factor. The sensitivity of the technique can be obtained from Eqs. (14) and (15) and are basically a function of the enzyme loading and the enzyme membrane thickness. Practical limits of detection vary considerably but a survey of the literature indicates that 10^{-5} M is not an unreasonable expectation. A recent publication has employed a trapping technique for enhancing the response of amperometric probes[68)].

A survey of the amperometric electrodes which have been reported is given in Table 7. We would be remiss, however, if it were not stated that commercial instrumentation based on immobilized enzyme technology is available. Both Yellowsprings Instrument Company and Radiometer have introduced glucose analyzers based on a glucose oxidase amperometric electrode system. The cost of the instruments is about $ 3,000. They typify the potential of immobilized enzyme instruments requiring less than 2 min per assay, a sample volume of less than 100 μl, and reagent costs of about 2 cents per assay.

4 Immobilized Enzyme Reactors

The second class of immobilized enzyme based methods of analysis involves the use of a reactor to convert the analyte (substrate) to product and a measurable species. Although a number of reactor types exist, only two have been of significance in analytical applications: packed or fixed bed reactors and open tubular reactors (OTHERS). Of all of the analytical application forms, these may be the most important due to their compatibility with continuous flow systems, their capacity for high rates of sample throughput, and the wide variety of detection systems with which they may be interfaced. The collection of detectors includes potentiometric, amperometric, photometric, fluorometric, chemiluminescent, and thermal methodology. Immobilized enzyme reactors (IMERs) are also useful in the equilibrium mode, where large quantities of enzyme can be used to totally convert the substrate to product during a single pass through the reactor. This mode of operation utilizes one of the major *advantages* of immobilized enzyme systems and avoids the complexities of kinetic analysis. For all of the above reasons, IMER applications have been rapidly increasing.

In any analytical system based on an immobilized enzyme reactor, three factors are involved in the success of the system: conversion of the substrate to product and a detectable species; mass dispersion in the flow stream which may act to diminish the signal; and detection of a species or physical property whose change is proportional to the substance of interest. In order to develop an optimized system, each of these parameters must be considered. The first of these, conversion of analyte, can be considered simply

Table 7. Amperometric enzyme probes

Substrate	Enzyme	Electrode	Response time	Linearity	Stability	Ref.
Alcohols	Alcohol oxidase	Pt (O_2) −0.6 V vs SCE	~2 min	Up to 5 mg% Ethanol	~120 d	79)
		Pt (H_2O_2) + 0.6 V vs SCE	~2 min	1–25 x 10^{-9} M Methanol	~24 h	80)
	Alcohol dehydrogenase	Pt (NADH)	–	0–1 M nonlinear	~8 samples	81)
L-amino acids	L-amino acid oxidase (E.C. 1.4.3.2)	Pt (H_2O_2) potential scan	30–60 s	1–400 x 10^{-6} M	10–12 d	82)
		Pt (O_2) −0.6 V vs SCE	1–2 min	0.1–1 x 10^{-3} M Phenylalanine	> 120 d	83)
Glucose	Glucose oxidase (E.C. 1.1.3.4)	Pt (H_2O_2) + 0.6 V vs SCE	1 min (steady state)	0.5–15 x 10^{-3} M	300 d	84)
		Pt	3–10 min	1 – 20 x 10^{-3} M	–	85)
		Clark O_2 electrode	–	0.1–2.0 mg l^{-1}	–	86)
Diamines	Diamine oxidase	Pt (O_2) −0.6 V vs SCE	< 15 s	20–400 x 10^{-6} M	25% decrease in 14 d	87)

Lactate	Lactate dehydrogenase (E.C. 1.1.2.3)	Pt ($Fe(CN)_6$) +0.25 V vs SCE	~1 min	0.1–5 x 10^{-3} M	–	88)
		Glassy carbon (NADH) +0.7 V vs SCE	3–4 min	0.1–4 x 10^{-3} M non-linear	Decrease of 20% in 2 h	89)
Uric acid	Uric acid oxidase	Pt (O_2) –0.6 V vs SCE	2–3 min	0.1–1 mg l^{-1}	~100 d	90)
Phosphate	Alkaline phosphatase/ glucose oxidase	Pt (O_2) –0.6 V vs SCE	1–2 min	1–10 x 10^{-3} M	~ 90 d	91)

by using the integrated Michaelis-Menten equation,

$$([S]_0 - [S]_t) + K_m \ln \frac{[S]_0}{[S]_t} = Vt\,, \tag{18}$$

where $[S]_0$ is the initial substrate concentration, $[S]_t$ is the concentration of substrate at time t, t is time, and the remaining symbols have their normal definition. If we now let X be the fractional conversion of substrate to product, i.e.,

$$X = \frac{[S]_0 - [S]_t}{[S]_0}\,, \tag{19}$$

then the equation can be written

$$X[S]_0 - K_m \ln(1 - X) = V \frac{\epsilon L}{u}\,. \tag{20}$$

There are three parameters which can be varied in developing an analysis: the fractional conversion, the V which is related to the amount of enzyme present, and the time during which the analysis is run. The time of analysis in a flow analyzer would, of course, be related to the porosity, ϵ, and the length, L, of the enzyme reactor and the flow rate, u. Several groups have verified the validity of this relation[92, 93]. Unfortunately, due to the effects of mass transfer and the effects of immobilization on the enzyme kinetics, the situation is much more complex than presented above. Since the design and operation of immobilized enzyme reactors is considered by Pitcher[94], we will limit our treatment of the subject to peculiarities associated with analytical applications. It should be borne in mind, however, that the major differences between IMERs for industrial processes and those for analytical purposes are the scale, both in size and in time.

The second factor, mass dispersion, is an important consideration in analytical IMER systems. The width of the signal pulse in such a system determines two parameters of interest: the sample throughput rate and the signal height. Since the theoretical treatment dispersion in packed beds is quite different from that in open tubes, this is most conveniently discussed in the appropriate section below.

The final factor is the method of detection and is the primary topic of the remainder of this chapter. Since the detection system is the only contact the analyst has with the analysis, the authenticity of the representation is critical. The fidelity of the concentration profile provided by the detector is a function of a number of parameters including mixing in the detector volume, slow mass or heat transfer (if a thermal device is used) to the detector, and slow response characteristics of the associated electronics. There are essentially three effects observed from these parameters: the peak maximum is reduced; the maxima is displaced in time; and the peak is broadened. Several authors have presented quantitative treatments of these parameters for chromatographic peaks[95–98], and the situation for IMER detectors is analogous. A detailed discussion of this topic is beyond the scope of this review and the interested reader is referred to the literature[4]. It is, however, important to recognize the role of the detector in a logical optimization of an analytical system.

4.1 Packed Bed Reactors

As mentioned above, the use of IMERs in analysis has grown extremely rapidly. Packed bed reactors, by virtue of their similarity to chromatographic systems, have been particularly widely accepted. This approach is useful in developing equilibrium type assays. The size of the reactor would, in this case, be determined by the volume of support necessary for complete conversion of substrate to product. A discussion of the factors which can be used to optimize these systems is presented below.

Kinetic Considerations. As mentioned above, the kinetic situation in a packed bed IMER is more complex than the simple relation given above. Analytically, the problems reviewed by Pitcher[94)], influence the operation and design of an IMER in several ways. For a low enzyme activity per unit mass of carrier, the interparticle mass transfer is unimportant. When higher carrier specific activities are prepared on smaller particles to decrease the column size and dispersion while maintaining total conversion, the interparticle (or external) mass transfer effects increase and may become significant. Since the interparticle mass transfer influences only the K_m, the main effect will be to increase the amount of enzyme required for any fractional conversion. In a kinetic analysis system, the effects would be considerably greater. The analytical ramifications of these problems have been discussed in more detail elsewhere[4)].

Column Dispersion. The importance of sample dispersion and dilution during its passage through the column and associated tubing is the result of the limitation they may place on sample throughput and on the limit of detection. The former limitation arises from the fact that it would be preferable that sequential samples do not significantly overlap, a situation analogous to peak resolution in chromatography. The sample dilution can also be approximated by chromatographic theory.

Subject to the limitations and assumptions of the plate theory of chromatography, an infinitely narrow sample input function, and instantaneous total conversion of substrate to product, the concentration of product eluting from the reactor will be given by

$$[P]_t = [P]_{max} \exp - \frac{N(t - t_e)^2}{2 t_e^2}, \tag{21}$$

where

$$[P]_{max} \equiv \frac{C_a V_a}{V_e} \sqrt{\frac{N}{2\pi}}, \tag{22}$$

and $[P]_t$ is the concentration of product at any time t, $[P]_{max}$ is the maximum concentration of product, t_e is the elution time, N is the number of "theoretical plates," C_a is the concentration of analyte in the sample, V_a is the injected volume, and V_e is the elution volume. The most important result of this relation is the relationship between the maximum product concentration and the sample volume (V_a), the magnitude of the dispersion factors (N), and the size of the reactor and associated tubing (V_e). A detailed

tributed to the fact that the liquid slugs lay down a thin film of solution along the entire surface of the tubing which is picked up by subsequent slugs. Additional dispersion will be introduced by tubing joints, dialyzers, heating baths, unsegmented stream sections (spectrophotometer, sampler), and the debubbling process.

Early investigations into the dispersion observed in continuous flow systems were empirical in nature and restricted to either the dispersion due to the segmented stream flow or to that due to the unsegmented stream between the debubbler and the flow cell[106–110]. In deriving a model for the symmetrical dispersion observed in tubes, three assumptions were made: 1) instantaneous mixing of the film and segments of liquid; 2) constant dimensions of the segments and the film; and 3) negligibly slow longitudinal diffusion in the film. An expression can then be derived for the concentration of a species in the kth segment (C_k) of a series of flowing segments:

$$\frac{C_k}{C} = \frac{e^{-q}q^k}{(k!)} . \tag{25}$$

C is the concentration of the species of interest in a single segment (segment 0) before any dispersion has taken place and q is given by

$$q = \frac{V_f}{V_s} = \frac{4\, d_f L}{L_s d_t} , \tag{26}$$

where V_f and V_s are the volume of the film deposited from a segment during its passage through the tube and the slug respectively. It has been shown that these volumes are related to the thickness of the film, d_f, the length of the tube, L, the length of the segment, L_s, and the diameter of the tubing, d_t. From consideration of Eq. (25), it should be apparent that the distribution of the analyte through the k segments is a Poisson distribution with its maximum at q and $q-1$. For large values of q, the distribution approximates a Gaussian function with σ^2 equal to q.

The situation in continuous flow analysis does not involve one sample segment but a series of sample segments followed by a series of "wash" segments. The distribution of analyte thus occurs at the edge of the sample slug series. Although these equations provide an empirical treatment of the dispersion process and its relation to tube length and diameter, and the slug length, no fundamental understanding of the dispersion process, a prerequisite for optimization of continuous flow systems, can be gleaned from this data. Snyder and Adler have recently treated this model in a more rigorous manner[108, 109].

As mentioned earlier, other factors affect the dispersion of sample. Coupling between various analyzer components, dialyzers, heating baths, and unsegmented stream sections contribute to dispersion but have not been treated theoretically. In addition, the above discussion deals only with straight lengths of tubing, ignoring the effects of coiling. It is well known that coiling increases radial mass transport but the treatment of such fluid flow is beyond the scope of this review.

Kinetic Considerations. Despite the potential applicability of open tubular enzyme reactors in many areas of analysis, several intuitive problems would seem to preclude their use. First, the amount of enzyme that can be immobilized on a section of tubing is limited due to the relatively small surface area. A 1 mm diameter tube will have a geometric surface area of 31.4 cm^2 m^{-1} as opposed to a surface area of several hundred square meters per gram of small porous support beads. Thus, reactor lengths might become prohibitively large, i.e., 200 m. Second, one might expect severe mass transfer problems as the substrate diffuses radially. This might result in even longer reactors to achieve any reasonable fractional conversion. Despite these limitations, Hornby and his coworkers were able to obtain reasonable conversion in relatively short (3–10 m) reactors[110–115]. As mentioned in the previous section, axial dispersion is not eliminated by segmentation of the stream and in fact is proportional to the length of tubing used. Thus, to optimize the throughput rate and sensitivity of this technique, a better understanding of the basic processes occurring in the flowing stream is required.

Horvath and his coworkers[116–118] have studied the effects of segmentation and coiling on the efficiency of tubular enzyme coils. These workers noted that the dominant mode of mass transport under their conditions is convection and recirculating flow as a result of the axial and radial bolus flow regimens. Two important parameters in the efficient use of this bolus flow are the size of the liquid segments and the relative diameter of the coil. This can be seen from Fig. 6. Engasser and Horvath have concluded

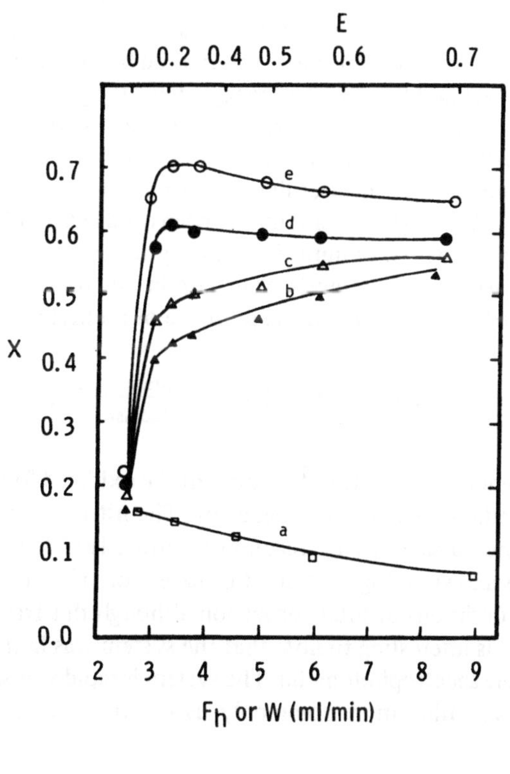

Fig. 6. Plots of conversion as a function of total flow rate for an open tubular reactor. (*a*) Represents non-segmented flow in a straight tube; (*b*) is a straight tube with segmented flow; (*c*), (*d*), and (*e*) represent coiled tubes with segmented flow with coil radius for (*c*) > (*d*) > (*e*) (reproduced from reference[116]) with permission)

that in fact open tubular reactors provide higher mass transfer rates than packed bed reactors at the same pressure drop[118].

If one is to take advantage of the high mass transfer rate and the benefits of mass transfer limited kinetics (vide supra), it becomes important to immobilize relatively large amounts of enzyme to the wall. Several approaches have been attempted to increase the active enzyme coupled. Direct coupling has been used in conjunction with a rigorous etching of the nylon surface[112, 119]. Alternately, a polymeric "fuzzy" wall can be attached to the nylon tube after which the polymer can be activated for enzyme coupling[120]. The third method involves the formation of an annulus of enzyme along the inner wall of the tube with either a polymer or silica[118]. The latter two approaches have resulted in reactors about 100 cm long which achieve reasonable conversions and are diffusion limited. It should be noted that with use of an annulus, product dispersion may exceed that anticipated due the diffusion in the annulus. Despite the limitations, the tremendous potential of tubular reactors in automated continuous flow analysis ensures their continued importance in immobilized enzyme analysis.

4.3 Analytical Applications of IMERs

The first successful automated analytical applications using immobilized enzymes was reported over a decade ago by Hicks and Updike[16]. Using polyacrylamide entrapped lactate dehydrogenase (LD) and glucose oxidase (GO), lactate and glucose were quantitated by spectrophotometric detection of a coupled dye. It is interesting to note that the linear dynamic range was limited by diffusion into the large particles that were used, a fact that these authors recognized. Using their automated system, up to 20 determinations per h could be performed. Since that time, over 50 papers have appeared that deal with packed bed IMERs. We will review a number of these applications with particular emphasis on the usefulness of the various detection systems.

Although colorimetry has been used in a number of applications, one which illustrates the potential of immobilized enzymes in trace analyses involves the determination of part per billion levels of nitrate in environmental water samples[121]. The nitrate was reduced to nitrite in the presence of methyl viologen, i.e.,

$$NO_3^- + 2\,H^+ + 2\,MV^+ \xrightarrow[\text{reductase}]{\text{nitrate}} 2\,MV^{2+} + NO_2^- + H_2O\,, \quad (27)$$

where MV^+ and MV^{2+} represent the oxidized form of methyl viologen and its one electron reduction product respectively. The nitrite produced in the reaction was determined by the classical azo dye reaction monitored at 543 nm. A schematic diagram of the system is shown in Fig. 7. A limit of detection of 17 ppb was postulated for the system under conditions of total conversion although this fractional conversion was never achieved. It is interesting to note that the system was limited by the flow stream noise and not the spectrophotometer. The system is rapid and simple and showed improved specificity over other more commonly used nitrate methods. These advantages were significant

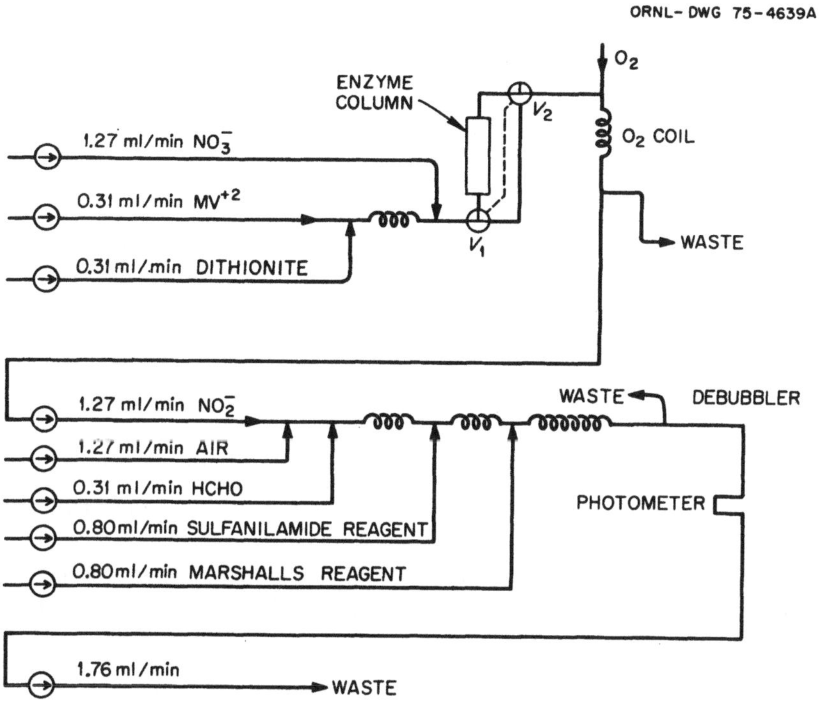

Fig. 7. Schematic diagram of a system for the determination of a nitrate in environmental water (reproduced from Ref.[121]) with permission of the American Chemical Society)

enough to cause the U.S. Environmental Protection Agency to choose it as an official analysis technique.

Recently, a system coupling a variable wavelength high performance liquid chromatography detector with a small IMER has been reported[122]. It would be expected from the earlier discussion that the small flow cell volume might result in improved limits of detection and a higher rate of sample throughput. An example of the data presented in the paper is shown in Fig. 8. From this data for the reduction of ethyl alcohol by alcohol dehydrogenase, it would appear that throughput rates of over 100 samples per hour are tenable. Using on-column sample injection and small (2–3 mm ID x 5 cm) reactors, relatively low limits of detection were obtained for glucose, ethanol, and testosterone. Using the LC detector, the limit of detection is 2.5×10^{-9} M (~2.5 mM) as opposed to a 1×10^{-8} M limit of detection with an 80 μl flow cell. Numerous other examples of both UV and visible photometry are reported in the literature.

Electrochemical transducers have also proven to be very useful as detectors for IMER systems. Both potentiometric and amperometric sensors have been used. Although a number of systems have been reported, all of the comercially available IMER devices are based on electrochemical methodology. Kimble division of Owens-Illinois have produced a system for blood urea measurements based on an ammonia ion selective electrode.

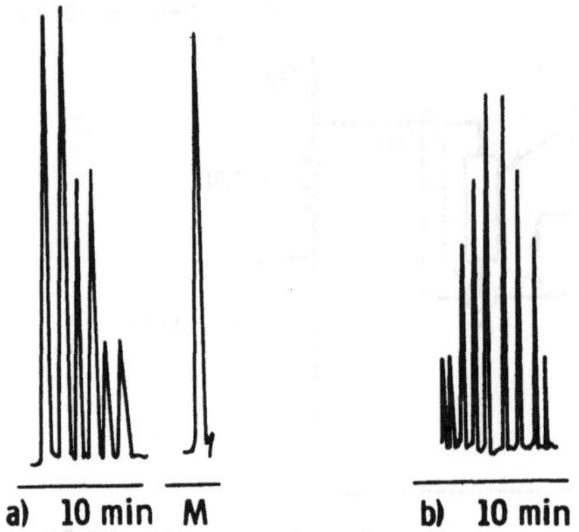

Fig. 8. Detector response for duplicate injections of glucose (a) and a diluted malt syrup (M) into a glucose oxidase/horse radish peroxidase IMER and of ethanol into an alcohol dehydrogenase IMER using a HPLC variable wavelength detector (reproduced from Ref.[122]) with permission)

The interesting feature of the system is the use of a packed bed IMER to obtain complete conversion of urea to ammonium ions. A schematic diagram of their instrument is shown in Fig. 9.

A second electrochemical system has been produced by Leeds and Northrup. Using amperometric detection downstream from a packed bed reactor, the Enzymax analyzer

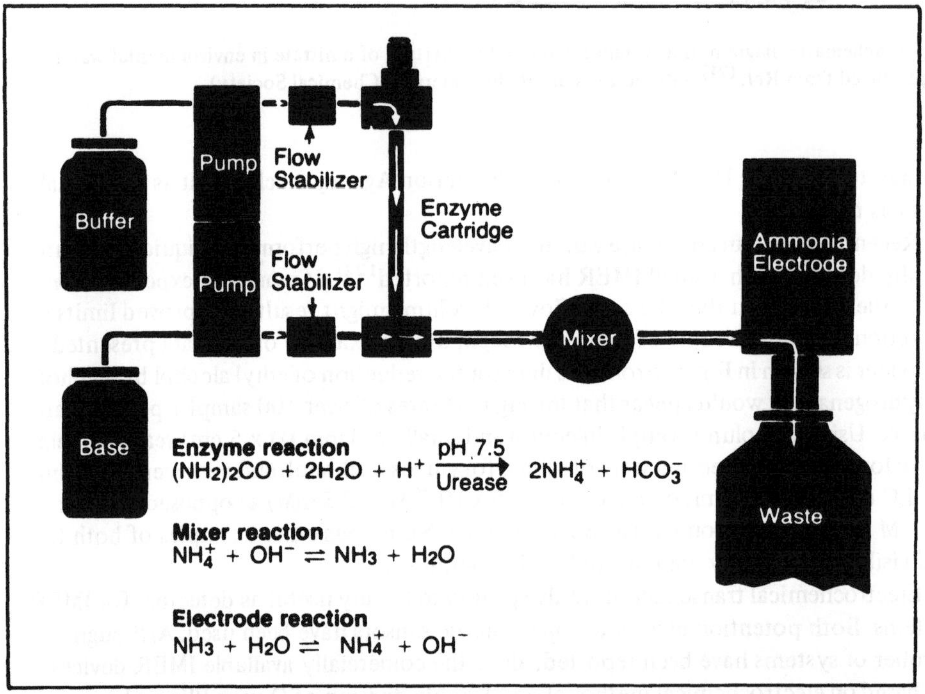

Fig. 9. Diagram of annion selective electrode IMER system for the analysis of urea in blood (reproduced from Ref.[134]) with permission)

in its simplest form is a single channel for glucose analysis. Dual channel analyzers are available for sucrose/glucose using immobilized invertase, mutarotase and glucose oxidase in one column and glucose oxidase in a parallel column and for lactose/glucose using immobilized β-galactosidase. Various components are available for connecting the basic analyzer to production lines for continuous on-line monitoring. The commercial system is based on amperometric glucose system reported by Weibel et al.[123].

One of the most popular areas of immobilized enzyme analyses has been the development of thermal analysis system. Early investigations centered around coupling IMERs to heat flow calorimeters[124]. These systems had excellent sensitivity but were very slow having sample analysis times of over 10 min per sample. Two groups have developed systems which seem to overcome the problems of thermal analysis[125, 126]. Our approach is shown in Fig. 10. For a urease packed bed reactor system, good sensitivity and correlation with accepted methods were observed along with sample throughput rates of over 60 samples per h[126]. The combination of the specificity of an immobilized enzyme reactor with a universal detector appears to be particularly appealing. A number of determinations can be visualized in which a reagentless system could be used instead of a coupled enzyme system or a less specific chemical reaction.

A number of other detection systems have been used including conductance and chemiluminescence. Due to spacial limitations, a compilation of detection systems are shown in Table 9.

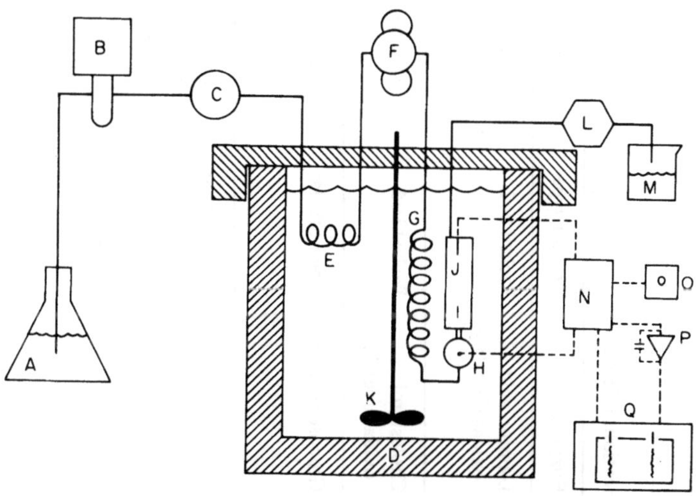

Fig. 10. Apparatus designed for flow enthalpimetric detection of serum constituents based on their specific degradation in an IMER. *A* reservoir; *B* pump; *C* pulse dampner-pressure gauge; *D* insulated water bath; *E* pre-equilibration coil; *F* sample injection valve; *G* equilibration coil; *H* reference termistor; *I* adiabatic column (IMER); *J* measuring thermistor; *K* stirrer (reproduced from Ref.[98] with permission)

Table 9. Analytical applications of IMERs

Analyte	Enzyme	Reactor type	Detection system	Comments	Ref.
L-amino acids	L-amino acid oxidase/catalase	Packed bed	Ammonia gas sensor	Reactor at 37 °C; Catalase used to recycle O_2 conversion; Linearity $3 \times 10^{-5} - 10^{-3}$ M	127)
Cholesterol	Cholesterol oxidase/catalase	Packed bed	Thermal	Linear 30–100 x 10^{-6} M Catalase thermistor	125)
Glucose	Glucose oxidase	Open tubular	Colorimetric	Detect I_2 color. Comparison of tubular, dialysis membrane, and column in continuous flow	
		Open tubular	Colorimetric/amperometric	Electrode linear to 2.5 x 10^{-3} M	
		Open tubular	Colorimetric	60 samples/h; 25,000 assays on a single tube	128)
		Packed bed	Amperometric	40/50 mesh titania support; 9 μl sample; linear to 5 g l^{-1}	129)
		Packed bed	Chemiluminescence	H_2O_2 – luminol – $Fe(CN)_6$ system	130)
		Packed bed	Differential conductivity	Co-immobilized glucose oxidase/catalase	135)
	Hexokinase/ glucose-6-phosphate dehydrogenase	Open tubular	UV Absorbance (340 nm)	Co-immobilized enzymes	131, 132)

Urea	Urease	Open tubular	Colorimetric	Chaney-Marbach method in Autoanalyzer system. Linear to at least 16 mg l^{-1}	133)
		Packed bed	Potentiometric		134)
		Packed bed	Thermal	Differential measurement across adiabatic column. 60 samples per h	126)
Uric acid	Uricase	Packed bed	Amperometric	End point method. Linear range $10–100 \times 10^{-6}$ M	136)
Penicillin	Penicillinase	Packed bed	Potentiometric		137)
Phosphate	Alkaline Phosphatase	Packed bed	Colorimetric	p-NO_2 phenylphosphate substrate inhibited by phosphate. Linear $10–100 \times 10^{-3}$ M	138)
Organic pesticides	Cholinesterase	Membrane reactor	Amperometric		139)

5 Miscellaneous Applications

Although enzyme electrodes and IMERs have received most of the attention in immobilized enzyme based analysis, several other types of applications have been presented. Solid surface fluorescence has been developed by Guilbault and his coworkers[19, 140–145]. In this technique, any enzymes and co-factors necessary to catalyze the formation of a fluorescent product are lyophilized onto a silicon rubber pad. When an analysis is required, the pad is reconstituted and the analyte solution added to the "immobilized enzyme" spot. Quantitation is accomplished by measuring the rate of production of fluorescence in a modified fluorometer. The only advantage that this technique has in common with other immobilized enzyme techniques is the increased reagent stability reported.

A second extremely versatile application using an immobilized enzyme is the immobilized enzyme stirrer. The prototype stirrer was simply a Teflon stir bar with immobilized urease held in place by a nylon net[146]. Used in conjunction with an ammonia ion selective electrode, initial response times and sensitivity were reported as 2 s and equivalent to 70% coversion. Similar systems have been reported for the fluorometric analysis of glucose[147] and the amperometric determination of cholesterol[148]. As might be expected, the conversion of substrate is a strong function of stirring rate. The authors noted that cavitation effects limited the ultimate stirring rate. Despite this problem, the glucose method had a c.v. of 1.3% in the range of clinical interest. It would appear that the immobilized enzyme stirrer can be used in a number of configurations and potentially replace any soluble enzyme assay. It should be noted, however, that the sample throughput is limited to 30 h^{-1} by wash out problems.

The potential use of enzymes in separations has recently been investigated. The use of urease immobilized inside of an anion exchange membrane allowed ammonium ion enrichment of a recipient solution[149]. A linear relationship was found between the concentration of urea in the donor solution and the ammonium enrichment over the range 10^{-5} to 10^{-1} M urea. It is interesting to note that the latter value exceeds the K_m of urease again indicating the advantage of mass transfer rate limitations. A definite contribution to the result was made by the anion exchange membrane itself. The potential of this technique as, for example, a tubular reactor-separator with down stream detection in the recipient stream might provide excellent methods for clinical and on-line industrial determinations.

6 Summary and Future Trends

It should be apparent from the preceding discussion that the use of immobilized enzymes in analytical applications is enjoying tremendous popularity. Until recently, the applications have been mostly demonstrations of feasibility. The principles of operation of many of the devices have been investigated and in some cases a model system provides

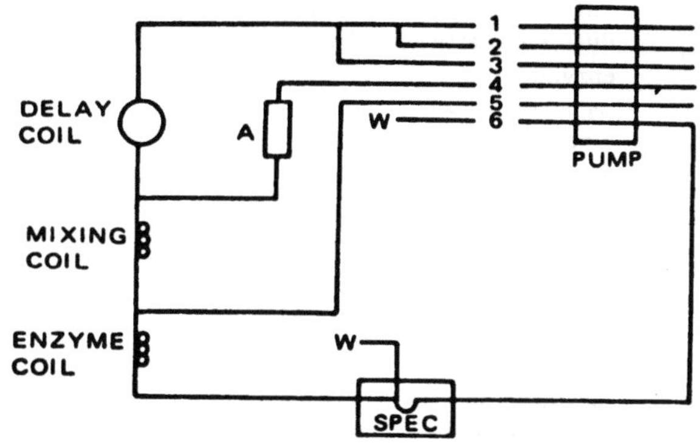

Fig. 11. Schematic diagram of continuous flow analyzer in which NADH is produced by an IMER. A packed bed of immobilized alcohol dehydrogenase is inserted at position A and a solution of NAD and ethanol pumped through the column producing the NADH required for the analytical reaction. The enzyme coil in the drawing is a tubular reactor of malate dehydrogenase required for the determination of glutamate oxaloacetate transaminase activity (reproduced from Ref.[113] courtesy of Plenum Press)

a good description of the observed behavior. This is an important step in the development of a new area of endeavor since subsequent efforts can be directed to optimization of the measurement system. We anticipate that this will be the direction of analytical applications in the next few years.

Another area which should receive increased attention is the development of reagentless analysis systems. Blaedel and Jenkins[89] have reported a reagentless lactate enzyme electrode in which both the enzyme and cofactor are retained within an outer membrane. The NADH produced by lactate dehydrogenase is reduced at a glassy carbon electrode and recycled; quantitation is achieved by measuring the steady-state current.

As pointed out earlier, immobilized enzymes are potentially a generating system for expensive or unstable reagents. It is generally found that reduced cofactors are more expensive than their oxidized counterparts. Hornby and his co-workers have used the system shown in Fig. 11 to generate NADH from NAD. A small reactor of alcohol dehydrogenase was supplied with a continuous flow of NAD and ethanol and the resultant NADH used in the analysis of the enzyme glutamate oxaloacetate transaminase. In this way, one of the major cost components of automated analysis can be diminished substantially.

7 Table of Symbols

C_k	concentration in k^{th} segment
d_f	film thickness
d_p	support particle diameter
d_t	tubing diameter

D_s	diffusion coefficient of substrate
D_p	diffusion coefficient of product
E	enzyme
k	segment number
k_2	kinetic constant for enzyme-substrate dissociation
K_m	Michaelis constant
L	column length
L_s	slug length
N	plate number
N_s	number of moles of substrate
P	product of enzyme reaction
r	column radius
S	substrate
t_e	elution time
u	flow velocity (cm s^{-1})
V_a	volume of sample
V_e	volume of analysis system
V_f	volume of film deposited
V_s	slug volume
$\bar{V}$	volume of the enzyme membrane
X	enzyme membrane thickness
ϕ_s	membrane permeability
ϵ	porosity of column
σ	variance of Gaussian peak

8 References

1. Weetall, H.H.: Anal. Chem. *46*, 602A (1974)
2. Guilbault, G.G.: Anal. Chem. *42*, 334, R (1970)
3. Bowers, L.D., Carr, P.W.: Anal. Chem. *48*, 544A (1976)
4. Bowers, L.D., Carr, P.W.: Immobilized enzymes in analytical and clinical chemistry: fundamentals & applications. New York: Wiley-Interscience (in press)
5. Mosbach, K. (ed.): Methods in enzymology, Vol. 44. New York: Academic Press 1976
6. Zaborsky, OR.: Immobilized enzymes. Cleveland: CRC Press 1973
7. Weetall, H.H. (ed.): Immobilized enzymes, antigens, antibodies, and peptides. New York: Marcel Dekker 1975
8. Pye, E.K., Wingard, L.B. (eds.): Enzyme engineering, Vol. II. New York: Plenum Press 1973
9. Means, G.E., Feeney, R.E.: Chemical modification of proteins. San Francisco: Holden-Day 1971
10. Zittle, C.A.: Adv. Enzymol. Relat. Areas Mol. Biol. *14*, 319 (1953)
11. Messing, R.A.: J. Non-Cryst. Solids *19*, 277 (1975)
12. Caldwell, K.D., Axen, R., Bergwall, M., Porath, J.: Biotech. Bioeng. *18*, 1573, 1589 (1976)
13. Sundaram, P.V., Crook, E.M.: Can. J. Biochem. *49*, 1388 (1971)
14. Thang, M.N., Graffe, M., Granberg-Manago, M.: Biochem. Biophys. Res. Commun. *31*, 1 (1968)
15. Chrambach, A., Rodbard, D.: Science *172*, 440 (1970)
16. Hicks, G.P., Updike, S.J.: Anal. Chem. *38*, 726 (1966)
17. Guilbault, G.G., Das, J.: Anal. Biochem. *33*, 341 (1970)
18. Blaedel, W.J., Kissel, T.R., Boguslaski, R.C.: Anal. Chem. *44*, 2030 (1972)
19. Lau, H.K., Guilbault, G.G.: Clin. Chem. *19*, 1045 (1973)

20. Johnson, P., Whately, T.L.: J. Colloid Interface Sci. *37*, 557 (1973)
21. Richards, F.M.: Ann. Rev. Biochem. *32*, 268 (1962)
22. Goldman, R., Silman, L.H., Caplan, S.R., Kedem, O., Katchalski, E.: Science *150*, 758 (1963)
23. Wold, F.J.: Biol. Chem. *236*, 106 (1961)
24. Manecke, G., Gunzel, G.: Naturwissenschaften *54*, 647 (1967)
25. Zaborsky, O.R.: Immobilized enzymes, p. 62. Cleveland: CRC Press 1974
26. Zahn, H., Meinenhoffer, H.: Makromol. Chem. *26*, 126 (1958)
27. Schick, A.F., Singer, S.J.: J. Biol. Chem. *236*, 2477 (1961)
28. Kitagawa, T., Aikawa, T.: J. Biochem. *79*, 233 (1976)
29. Guire, P., Fliger, D., Hodgson, J.: Pharmaco. Res. Commun. *9*, 121 (1977)
30. Zaugg, R.H., King, L., Klotz, I.M.: Biochem. Biophys. Res. Commun. *64*, 1192 (1975)
31. Inman, J.K., Dintzis, H.M.: Biochemistry *8*, 4074 (1969)
32. Levin, Y., Pecht, M., Goldstein, L., Katchalski, E.: Biochemistry *3*, 1905 (1964)
33. Manecke, G., Singer, S.: Makromol. Chem. *39*, 13 (1960)
34. Bar-Eli, A., Katchalski, E.: Nature *188*, 856 (1960)
35. Grubhofer, N., Schleith, L.: Hoppe-Seyler's Z. Physiol. Chem. *297*, 108 (1954)
36. Biodi, P.A., Pace, M., Brenna, O., Pietta, P.G.: Eur. J. Biochem. *61*, 171 (1976)
37. Sundaram, P.V., Hornby, W.E.: FEBS Letts *10*, 325 (1970)
38. Goldstein, L., Pecht, M., Blumberg, S., Atlas, D., Levin, Y.: Biochemistry *9*, 2322 (1970)
39. Axen, R., Porath, J., Ernback, S.: Nature *215*, 1491 (1967)
40. Weetall, H.H., Hersch, L.S.: Biochim. Biophys. Acta *185*, 464 (1969)
41. Weetall, H.H., Detar, C.C.: Biotech. Bioeng. *17*, 295 (1975)
42. Coulet, P.R., Julliard, J.H., Gautheron, D.C.: Biotech. Bioeng. *16*, 1055 (1974)
43. Robinson, P.J., Dunnill, P., Lilly, M.D.: Biotech. Bioeng. *15*, 603 (1973)
44. Zaborsky, O.R., Ogletree, J.: Biochem. Biophys. Res. Commun. *61*, 210 (1974)
45. Zaborsky, O.R.: In: Enzyme engineering. Pye, E.K., Wingard, L.B. (eds.), Vol. II, p. 115. New York: Plenum Press 1974
46. Goldstein, L.: Biochemistry *11*, 4072 (1972)
47. Goldstein, L., Katchalski, E.: Z. Anal. Chem. *243*, 375 (1968)
48. Johansson, A.C., Mosbach, K.: Biochim. Biophys. Acta *370*, 348 (1974)
49. Weetall, H.H.: Biochim. Biophys. Acta *212*, 1 (1970)
50. Dixon, J.D., Stolzenbach, F.E., Berenson, J.A., Kaplan, N.O.: Biochem. Biophys. Res. Commun. *52*, 905 (1973)
51. Messing, R.A.: In: Adv. Biochem. Engineering. Ghose, T.K., Fiechter, A., Blakebrough, N. (eds.), Vol. 10, p. 51. Berlin, Heidelberg, New York: Springer Verlag 1978
52. Cooney, C.L., Weaver, J.C., Tannenbaum, S.R., Faller, D.V., Shields, A., Jahnke, M.: In: Enzyme engineering. Pye, E.K., Wingard, L.B. (eds.), Vol. II, p. 411. New York: Plenum Press 1974
53. Weaver, J.C., Cooney, C.L., Fulton, S.P., Schuler, P., Tannenbaum, S.R.: Biochim. Biophys. Acta *452*, 285 (1976)
54. Racine, P., Mindt, W.: Experentia Suppl. *18*, 525 (1971)
55. Blaedel, W.J., Boguslaski, R.C.: Biochem. Biophys. Res. Commun. *47*, 248 (1972)
56. Llenado, R.A., Rechnitz, G.A.: Anal. Chem. *43*, 1457 (1971)
57. Guilbault, G.G., Montalvo, J.G.: J. Am. Chem. Soc. *92*, 2533 (1970)
58. Cullen, L.F., Rusling, J.F., Schleifer, A., Papariello, G.J.: Anal. Chem. *46*, 1955 (1974)
59. Papastathopoulos, D.S., Rechnitz, G.A.: Anal. Chem. *48*, 862 (1976)
60. Tran-Minh, C., Broun, G.: Anal. Chem. *47*, 1359 (1975)
61. Guilbault, G.G., Tarp, M.: Anal. Chim. Acta *73*, 355 (1974)
62. Nilsson, H., Akerlund, A., Mosbach, K.: Biochim. Biophys. Acta *320*, 529 (1973)
63. Montalvo, J.G., Guilbault, G.G.: Anal. Chem. *41*, 1897 (1969)
64. Mascini, M., Liberti, A.: Anal. Chim. Acta *68*, 177 (1974)
65. Carr, P.W.: Anal. Chem. *49*, 799 (1977)
66. Guilbault, G.G.: Handbook of enzymatic methods of analysis, p. 445. New York: Marcel Dekker 1976

67. Mell, L.D., Maloy, J.T.: Anal. Chem. *47*, 299 (1975)
68. Mell, L.D., Maloy, J.T.: Anal. Chem. *48*, 1597 (1976)
69. Guilbault, G.G., Montalvo, J.G., Jr.: J. Am. Chem. Soc. *91*, 2164 (1969)
70. Guilbault, G.G., Hrabankova, E.: Anal. Chim. Acta *52*, 287 (1970)
71. Guilbault, G.G., Shu, F.R.: Anal. Chem. *44*, 2161 (1972)
72. Guilbault, G.G., Nagy, G.: Anal. Chem. *45*, 417 (1973)
73. Anfalt, T., Graneli, A., Jagner, D.: Anal. Letts. *6*, 969 (1973)
74. Nagy, G., Von Storp, L.H., Guilbault, G.G.: Anal. Chim. Acta *66*, 443 (1973)
75. Rechnitz, G.A., Llenado, R.: Anal. Chem. *43*, 283 (1971)
76. Meyerhoff, M., Rechnitz, G.A.: Anal. Chim. Acta *85*, 277 (1976)
77. Guilbault, G.G., Hrabankova, E.: Anal. Chem. *42*, 1779 (1970)
78. Wawro, R., Rechnitz, G.A.: J. Mem. Sci. *1*, 143 (1976)
79. Nanjo, M., Guilbault, G.G.: Anal. Chim. Acta *75*, 169 (1975)
80. Clark, L.C., Jr.: Biotech. Bioeng. Symp. No. 3, 377 (1972)
81. Suzuki, S., Takahashi, F., Satoh, I., Sonobe, N.: Bull. Chem. Soc. Japan *48*, 3246 (1975)
82. Nagy, G., Pungor, E.: Hung. Sci. Instrum. *32*, 1 (1975)
83. Nanjo, M., Guilbault, G.G.: Anal. Chim. Acta *73*, 367 (1974)
84. Guilbault, G.G., Lubrano, G.J.: Anal. Chim. Acta *64*, 439 (1973)
85. Williams, D.L., Doig, A.R., Jr., Korosi, A.: Anal. Chem. *42*, 118 (1970)
86. Notin, M., Guillien, R., Nabet, P.: Ann. Biol. Clin. *30*, 193 (1972)
87. Toul, Z., Macholan, L.: Coll. Czech. Chem. Comm. *40*, 2208 (1975)
88. Durliat, H., Comtat, M., Mohenc, J., Baudras, A.: J. Electroanal. Chem. *66*, 73 (1975)
89. Blaedel, W.J., Jenkins, R.A.: Anal. Chem. *48*, 1240 (1976)
90. Nanjo, M., Guilbault, G.G.: Anal. Chem. *46*, 1769 (1974)
91. Guilbault, G.G., Nanjo, M.: Anal. Chim. Acta *78*, 69 (1975)
92. Wilson, R.J.H., Kay, G., Lilly, M.D.: Biochem. J. *108*, 845 (1968)
93. Havewala, N.B., Pitcher, W.H.: In: Enzyme engineering. Pye, E.K., Wingard, L.B. (eds.), Vol. II, p. 315. New York: Plenum Press 1974
94. Pitcher, W.H.: In: Adv. Biochem. Engineering. Ghose, T.K., Fiechter, A., Blakebrough, N. (eds.), Vol. 10, p. 1. Berlin, Heidelberg, New York: Springer 1978
95. Sternberg, J.C.: In: Advances in gas chromatography. Giddings, J.C., Keller, R.A. (eds.), Vol. 2, p. 206. New York: Marcel Dekker 1966
96. Littlewood, A.B.: Gas chromatography. New York: Academic Press 1970
97. Johnson, H.W., Stross, F.H.: Anal. Chem. *31*, 357 (1959)
98. Schifreen, R.S., Hanna, D.A., Bowers, L.D., Carr, P.W.: Anal. Chem. *49*, 1929 (1977)
99. Giddings, J.C.: Dynamics of chromatography, Part I. New York: Marcel Dekker 1965
100. Kirkland, J.J. (ed.): Modern practice of liquid chromatography. New York: Academic Press 1971
101. Karger, B.L., Martin, M., Guiochon, G.: Anal. Chem. *46*, 1640 (1974)
102. Skeggs, L.J.: Am. J. Clin. Path. *28*, 311 (1957)
103. Thiers, R., Oglesby, K.: Clin. Chem. *10*, 246 (1964)
104. Thiers, R., Oglesby, K., Kirsch, W.: Clin. Chem. *13*, 451 (1967)
105. Thiers, R., Jeyn, M., Wildermann: Clin. Chem. *16*, 832 (1970)
106. Walker, W.H.C., Pennock, C.A., McGowan, G.K.: Clin. Chim. Acta *27*, 421 (1970)
107. Walker, W.H.C.: In: Continuous flow analysis. Furman, W.B. (ed.), p. 207. New York: Marcel Dekker 1975
108. Snyder, L.R., Adler, H.J.: Anal. Chem. *48*, 1017 (1976)
109. Snyder, L.R., Adler, H.J.: Anal. Chem. *48*, 1022 (1976)
110. Hornby, W.E., Filippusson, H., MacDonald, A.: FEBS Lett. *9*, 8 (1970)
111. Inman, D.J., Hornby, W.E.: Biochem. J. *129*, 255 (1972)
112. Campbell, J., Hornby, W.E., Morris, D.L.: Biochim. Biophys. Acta *384*, 307 (1975)
113. Hornby, W.E., Campbell, J., Inman, D.J., Morris, D.L.: In: Enzyme engineering. Pye, E.K., Wingard, L.B. (eds.), Vol. II, p. 401. New York: Plenum Press 1974

114. Inman, D.J., Hornby, W.E.: Biochem. J. *137*, 25 (1974)
115. Filippusson, H., Hornby, W.E., MacDonald, A.: FEBS Lett. *20*, 291 (1972)
116. Horvath, C., Sardi, A., Solomon, B.A.: Physiol. Chem. Phys. *4*, 125 (1972)
117. Horvath, C., Solomon, B.A., Engasser, J.-M.: Ind. Eng. Chem. Fundam. *12*, 431 (1973)
118. Horvath, C., Solomon, B.A.: Biotech. Bioeng. *14*, 885 (1972)
119. Morris, D.L., Campbell, J., Hornby, W.E.: Biochem. J. *143*, 497 (1974)
120. Sundaram, P.V., Apps, D.K.: Biochem. J. *161*, 441 (1975)
121. Senn, D.R., Carr, P.W., Klatt, L.N.: Anal. Chem. *48*, 954 (1976)
122. Cremonesi, P., Bovara, R.: Biotech. Bioeng. *18*, 1487 (1976)
123. Weibel, M.K., Dritschilo, W., Bright, H.J., Humphrey, A.E.: Anal. Biochem. *52*, 402 (1973)
124. Johannson, A.: Protides Biol. Fluids, Proc. Colloq. *20*, 567 (1973)
125. Mattiasson, B., Danielsson, B., Mosbach, K.: Anal. Letts. *9*, 217 (1976)
126. Bowers, L.D., Canning, L.M., Sayers, C.N., Carr, P.W.: Clin. Chem. *22*, 1314 (1976)
127. Johannson, G., Edstrom, K., Ogren, L.: Anal. Chim. Acta *85*, 55 (1976)
128. Léon, L.P., Narayanan, S., Dellenbach, R., Horvath, C.: Clin. Chem. *22*, 1017 (1976)
129. O'Driscoll, K.F., Kapoulas, A., Albisser, A.M., Gander, R.: In: Hydrogels for medical and related applications. Andrade, J.D. (ed.), ACS Symposium Series, No. 31, 1976
130. Bostick, D.T., Hercules, D.M.: Anal. Chem. *47*, 447 (1975)
131. Morris, D.L., Campbell, J., Hornby, W.E.: Biochem. J. *147*, 593 (1975)
132. Leon, L.P., Sansur, M., Snyder, L.R., Horvath, C.: Clin. Chem. *23*, 1556 (1977)
133. James, D.R., Pring, B.: Clin. Chim. Acta *62*, 435 (1975)
134. Watson, B., Keyes, M.H.: Anal. Letts. *9*, 217 (1976)
135. Messing, R.A.: Biotech. Bioeng. *16*, 897 (1974)
136. Dritschilo, W., Weibel, M.K.: Biochem. Med. *11*, 242 (1974)
137. Rusling, J.F., Luttrell, G.H., Cullen, L.F., Papariello, G.J.: Anal. Chem. *48*, 1211 (1976)
138. Weetall, H.H., Jacobson, M.A.: Proc. IV IFS Ferment. Technol. Today, 361 (1972)
139. Goodson, L.H., Jacobs, W.B.: Midwest Res. Inst. Reprint No. MRI 1173
140. Guilbault, G.G., Vaughan: Anal. Chim. Acta *55*, 107 (1971)
141. Guilbault, G.G., Zimmerman, R.L.: Anal. Letts. *3*, 133 (1970)
142. Reitz, B., Guilbault, G.G.: Clin. Chem. *21*, 1991 (1975)
143. Kiang, S.W., Kuan, J.W., Kuan, S.S., Guilbault, G.G.: Clin. Chem. *21*, 1799 (1975)
144. Reitz, B., Guilbault, G.G.: Clin. Chem. *21*, 715 (1975)
145. Reitz, B., Guilbault, G.G.: Anal. Chim. Acta *58*, 75 (1975)
146. Guilbault, G.G., Stokbro, W.: Anal. Chim. Acta *76*, 237 (1975)
147. Kiang, S.W., Kuan, J.W., Kuan, S.S., Guilbault, G.G.: Clin. Chem. *22*, 1378 (1976)
148. Huang, H., Kuan, S.S., Guilbault, G.G.: Clin. Chem. *23*, 671 (1977)
149. Blaedel, W.J., Kissel, T.R.: Anal. Chem. *47*, 1602 (1975)

MIX
Papier aus verantwortungsvollen Quellen
Paper from responsible sources
FSC® C105338

If you have any concerns about our products,
you can contact us on
ProductSafety@springernature.com

In case Publisher is established outside the EU,
the EU authorized representative is:
Springer Nature Customer Service Center GmbH
Europaplatz 3, 69115 Heidelberg, Germany

Printed by Libri Plureos GmbH
in Hamburg, Germany